3800 16 0002207 7
D0548570

The Apple Orchard

The Apple Orchard

The Story of Our Most English Fruit

PETE BROWN

PARTICULAR BOOKS
an imprint of
PENGUIN BOOKS

PARTICULAR BOOKS

UK | USA | Canada | Ireland | Australia
India | New Zealand | South Africa

Particular Books is part of the Penguin Random House group of companies
whose addresses can be found at global.penguinrandomhouse.com.

First published 2016
003

Set in 12.5/15.5 pt Baskerville MT Std
Typeset by Jouve (UK), Milton Keynes
Printed in Great Britain by Clays Ltd, St Ives plc

A CIP catalogue record for this book is available from the British Library

ISBN: 978-1-846-14883-5

For Liz,
the apple of mine eye

‘It is remarkable how closely the history of the apple tree is connected with that of man.’

Henry David Thoreau, *Wild Apples*, 1862

‘Why do we need so many different kinds of apples? Because there are so many different folks . . . There is merit in variety itself. It provides more points of contact with life, and leads away from uniformity and monotony.’

Liberty Hyde Bailey, *The Apple-Tree: The Open Country*, 1922

‘A is for Apple.’

Traditional

Contents

PART FOUR

Harvesting

PART FIVE

Celebrating

PART SIX

Transforming

PART SEVEN

Slumbering

Preface

Hecks' Orchard

We park on a bend in a narrow lane and walk down a grassy, rutted track hemmed in by rambling hedgerows. Cows eye us drily from the next field. Beyond them, the land curves away gently into suggestions of hills that are vaguely apologetic, as only English hills can be. The rattle of a tractor echoes from their direction. There's birdsong, the odd plane, the occasional rustle and soft whump of falling fruit. And, I swear, though it's probably imagined, a steam train in the distance.

As we near the gate, three stout old ladies in walking boots and armed with staffs tramp past and smile at us. 'The fruit is very good this year!' exclaims their leader.

As a city dweller, I dig out my own walking boots only for music festivals. I feel virtuous now I'm wearing them for something a little closer to their intended purpose than watching gloomy guitar bands in the pouring summer rain. As if in approval of my new mission, the dew of autumn's first chill has washed the boots clean of mud, and they're refreshingly light on my feet. Instead of being sucked at greedily by midnight swamps, they're swishing through jewelled grass that has been left to grow long to cushion the imminent fall. Instead of the snap and crackle of discarded beer cups being trodden into the mire, the sound from beneath my soles is the squeak of wet apple skins, followed occasionally by a juddering crunch as the fallen fruit is mulched into the turf.

The trees stretch before us in lines, a jumble of shape and colour. Until now, I'd never noticed how diverse apples can be. Some are no bigger than plums, others are the size of small grapefruit. The Victorians prized thin-skinned apples that seemed to melt in your mouth over the tougher, waxier skins we see here. Beneath that skin, the flesh of one may be dry and crumbly, while the next might be creamy and soft, or juicy and crisp. Some hang in bunches like grapes, while others – no more than two inches across – line spindly, delicate branches, fizzing out from the heart of the tree, like hair full of static electricity. And if I always thought of apples as simply green or red, I see now the whole spectrum of pink, blush, purple and rose. Some are as dark and rich as Bordeaux wine. Others are streaked – green with dark purple stripes like battle scars.

An orchard is not a field. It's not a forest or a copse. It couldn't occur naturally; it's definitely cultivated. But an orchard like this doesn't override the natural order: it enhances it, dresses it up. It demonstrates that man and nature together can – just occasionally – create something more beautiful and literally more fruitful than either could alone. The vivid brightness of the laden trees, studded with jewels, stirs some deep genetic memory and makes the heart leap. Here is bounty and excitement. Three months early, here is Christmas, replete with shining baubles.

The main track through the centre of the orchard is stained ferrous orange by crushed, oxidized apple flesh. Buzzing wasps, drunk on the fallen fruit, carve random paths through the air. The tree bark is woodpecker-scarred and encrusted with lichen. Where branches have been cut away to maximize exposure to the sun for those that remain, the boles left behind are home to vast, multicultural insect nations.

The landscape around the orchard almost demands myth-making from those who observe it. If the Arthurian legend didn't exist, you'd make it up on the spot as soon as you saw the unnatural majesty of Glastonbury Tor, which – let's be honest – is probably exactly what someone did. Reputedly the site of Avalon, also known as Avallach, the 'Isle of Apples', ruled by the faerie queen Morgan le Fay, this part of Somerset is the land of fairies and the dead. The name 'Avalon' derives from the Celtic prefix *av* or *af*, which means 'apple'. The Celts credited apples with the power of healing and youth. Merlin sat beneath an apple tree to teach, and Avalon is where King Arthur was brought to be healed, to the home of the fabled golden apples of immortality.

My companion here in Hecks' Orchard has been studying apples for the last ten years, and he's brought me here to show me why. Bill Bradshaw is a professional photographer who found that when he started taking pictures of apples and orchards he couldn't stop. Now he plucks one and takes a bite. 'This is a Kingston Black,' he says, holding it out to me. 'It's the king of cider apples. Perfectly balanced. It's bitter and sharp and dry and sweet all in one. Have a bite – you'll see what I mean.'

I do as I'm told. The first sensation in my mouth is one of perfect juice. As Bill says, it's bitter and dry and sweet all at once, running over my teeth and gums, all goodness and refreshment.

The second sensation is a buzzing on my tongue. This buzz quickly becomes more jarring, until it feels like I've been licking a battery. My tongue is too large for my mouth. Did I say licking a battery? It now feels as though someone is pumping a low-voltage electrical current through my gums, and not in a nice way.

After a couple of minutes, my throat says, 'Oh, you think your tongue is having problems?' What starts as an itch at the top of my throat soon becomes a swelling, a constriction. I'm swallowing frequently now, and each time I try, it's harder to do so. There's a blockage that wasn't there three minutes ago.

I can still breathe. I don't think I'm going to require Bill to perform an emergency tracheotomy with the pen I was using to take notes until a few moments ago. Nevertheless, after admitting one bite of apple, it's quite clear that my body has erected a hasty picket line to prevent entry of any further morsels.

I've experienced this reaction once or twice before, but I always imagined it was down to pesticides or other sprays, and it was never as bad as this – although now I think of it, it has become steadily more severe each time it has happened. Now, in the middle of a fully organic orchard, eating one tiny mouthful, I have to accept that I have developed a serious allergy to apples.

I ate apples perfectly happily while growing up. I never pushed a bag of crisps out of the way to get to one, but they were fine – juicy and satisfying, but quite monotone: the crowd-pleasing Golden Delicious that always made me wonder if toffee-apple-makers had got the relative proportions of toffee and apple the wrong way round; or the fat, watery culinary fruit that went into apple pies which, for me, were just an excuse to eat custard, because the school dinner ladies looked at you funny if you asked for a bowl of that on its own.

Now the apple has tricked me. After all these years of indifference, it has made me want it, desire it. A whole array of exotic riches, treasure growing on trees, promising a breadth of flavour sensations I could previously never have dreamed of. I gave in, and now I can never submit again: this new object of desire has been taken away from me even as it hangs in front of me.

For me, the apple really is the forbidden fruit.

PART ONE
Blossoming

April–May

'It was from out of the rind of one apple tasted, that the knowledge of good and evil, as two twins cleaving together, leaped forth into the world.'

John Milton, *Areopagitica*, 1644

1.

How Apples Work

About ten years ago, my wife, Liz, and I were on holiday in the French Pyrenees. In late summer, we climbed winding roads through forests and past *auberges* selling ham baguettes and chestnut beer, and even though we knew we were heading to some kind of summit, every time we reached it and were presented with snow-capped mountains standing stark and impossibly solid against a clear blue sky, the clouds at their feet rather than at their peaks, the simple reality of them was almost impossible to process.

But in this part of the world, some of the most extraordinary sights are beneath the mountains. The Pyrenees are dotted with cave complexes in which the effects of Stone Age life have been remarkably preserved. When we think of the Palaeolithic 'caveman', our impressions are framed largely by what's been found in sites like these.

Modern archaeological discoveries have questioned the notion that our ancestors ever really lived in caves, suggesting they were used only in winter months. While caves offer the perfect environment for the preservation of Stone Age artefacts, they give us only part of the story, which means we can easily get that story wrong.

There's a big difference between the evolution of *Homo sapiens* and our progress towards our current state. That may sound obvious, but we often forget this when we think about our ancestors. The time-worn depiction of the caveman is of a primitive brute, hitting anything that moves with his club, attempting to seduce his cavewoman with a succession of grunts before giving up and dragging her by the hair back to his dank, smelly lair. But ten, twenty, thirty thousand years ago, human brains were just as big and developed as they are now. We had no library of centuries-old accumulated knowledge, but we surely had ideas, judgements, emotions, a sense of humour, a sense of spirituality, and theories about how the world around us worked.

When Liz and I visited *la grotte de l'Ariège* in the Pyrenees, such thoughts came alive around us. Archaeologists now believe places such as this had some special symbolic or religious role rather than simply being everyday living spaces. The deeper we went under the mountain into the heart of the complex, where the paintings are densest, the less likely it seemed that they were mere decoration for Stone Age living-room walls. The caves with the finest paintings would not have been as comfortable or practical as those we passed through to get to them. The locations of some of the most impressive paintings suggest the artist endured long hours in uncomfortable or even dangerous positions.

Two things struck me about the art itself: the sheer quality and talent of the artist in rendering animal shapes that seem to move with lithe grace across the undulating cave walls, and the presence of abstract shapes whose meanings are still unknown. Are they mere doodles, or some kind of code or language? Either way, when you stand and look at the paintings, especially the hand silhouettes created by spitting paint from the mouth on to splayed fingers, you feel the presence of other human beings long gone: not club-wielding cavemen, but people just like us, as if they existed only moments ago.

The paintings I looked at were 14,000 years old. After researching similar paintings in the Chauvet Cave in southern France that are thought to be around 40,000 years old, archaeologist Steven Mithen wrote that these 'first representational paintings we have . . . are as technically accomplished and expressive as any painting by humankind – there was no gradual, cumulative evolution of the capacity for art'.

In his book *The Alchemy of Culture*, which explores the use of intoxicants in ancient societies, Richard Rudgley proposes that cave paintings of animals are more than simply a diary of what's been eaten, or a how-to guide to hunting. A study of the animals depicted on cave walls in Lascaux, south-western France, shows that only one reindeer features on the walls, yet the large quantity of reindeer bones in the debris on the cave floor suggests it was a dietary staple. There's a discrepancy between the animals painted and the animals that were hunted and eaten, leading researchers to conclude that the paintings had some symbolic significance that, so far, we can only guess at.

Five or six years after my visit to the caves, I was standing in an orchard in Somerset, my throat angry and swollen after my last ever bite of an apple. Behind my acute discomfort, I was feeling a rush of excitement from the tide of images and associations that had just flooded me. The apple as a symbol, a metaphor or badge, has always lurked at the edge of my subconscious, subliminally, as I'm sure it has for most people who live in the temperate climates where it thrives. I wasn't recognizing some hitherto unnoticed close affinity with the fruit – far from it – but the sudden rush of impressions and emotions, the deep stirring I felt in my soul, was an awakening, a realization of just how ubiquitous the apple is, and how little thought I'd given it until now.

We've done the same thing to the apple as our ancestors did with the animals they chose to portray on their sacred walls. Sure, apples have always been important to our diets: archaeological evidence of wild apples being collected and eaten in Europe goes back 11,000 years, and apples have been cultivated since at least 2000 BC. But while apple cultivation has been widespread across Europe and Asia for millennia, important as it is, the apple is hardly the most crucial ingredient of our diet. It's one fruit among many. And if our veneration had any relationship to real-world, practical importance, our myths and legends would surely be full of references to wheat instead. Hunter-gatherers first settled down to build permanent settlements so they could grow crops. Some form of grain – to bake bread or brew beer – is far more pervasive, and has remained more central to our diets throughout the history of civilization than the apple ever has.

Yet it's the apple that has been granted a symbolic role unmatched by anything else, across cultures and continents. As well as Avalon, there are the golden apples of

Greek myth, and silver and gold apples handed down through countless generations of fairy tales. Isaac Newton observed an apple falling to the ground* in a straight line in his mother's garden in Woolsthorpe Manor in Lincolnshire, inspiring him to formulate his theory of gravity. The first logo for Apple Computer, Inc. showed Newton sitting beneath a tree with the apple above him, about to fall. This was later changed to the rainbow-coloured apple with a bite taken from it, suggesting the acquisition of knowledge, and the Mac is named after McIntosh, a North American apple variety popular in school lunchboxes. There's the bright green Granny Smith of the Beatles' record label, which was named, according to Paul McCartney, because 'We [were] starting a brand-new form of business. So, what is the first thing that a child is taught when he begins to grow up? A is for Apple.' The story of William Tell shooting an apple from his son's head with a crossbow is merely one version of a legend that keeps cropping up in Germanic folklore through the centuries. There are enough different stories about how New York came to be known as 'The Big Apple' to fill a book of their own. Then there's the apple for teacher, the bad apple that spoils the bunch and the apple of my eye, who often upsets the apple cart. 'An apple a day keeps the doctor away' may sound like another one of these many proverbs, but it was actually coined by J. T. Stinson, a Missouri fruit specialist, at the 1904 St Louis World Fair in response to a sharp decline in the cultivation of apples caused by the temperance movement. Until the late nineteenth century,

* Not on his head, as popular myth often tells it.

America's apples were mainly cultivated for cider rather than eating, which was then a novel concept. Stinson saw the way things were going, and he wasn't just coming up with a sound bite – apples contain fibre, vitamins and flavonoids that play an important role in preventing many types of disease and promoting good digestion.

And then there's the most famous apple in the Western world, different from all the others in that it didn't give us the theory of gravity, or *Sergeant Pepper* or a more intuitive desktop interface, or any of the other revolutionary discoveries that have made our lives better: it gave us suffering, the experience of fruitless toil and agonizing labour, after precipitating our ejection from Paradise.

The size, shape, colour and texture of the apple all play their role in making it so symbolically and culturally important. But so does its ubiquity. We've domesticated the apple far more than we have most fruits. The apple tree is as much man's best friend as the dog is. Liberty Hyde Bailey was an American horticulturalist and botanist who co-founded the American Society for Horticultural Science in 1903. He wrote sixty-five books on horticulture, including *The Apple-Tree* in 1922. Despite the incredible breadth of his knowledge and activities, nothing matched his fascination with the apple:

> The apple-trees are of human habitations and human labor; they cluster about the buildings, or stand guard at a gate; they are in plantations made by hands. As I see them again, I wonder whether any other plant is so characteristically a home-tree. So is the apple-tree, even when full grown, within

> the reach of children. It can be climbed. Little swings are hung from the branches. Its shade is low and familiar. It bestows its fruit liberally to all alike.

This familiarity hides the apple tree in plain sight. Like air or water or beer, it's so much a part of the everyday, we forget how special it is.

I was in Hecks' Orchard with Bill Bradshaw that day because we were writing a book about cider together. Over the next two years, we explored orchards around the world, and when we finished that book I was left with a fascination for orchards and apples that went far beyond the drink they make. I wanted to explore the apple in all its guises, and I found that I missed orchards and had a yearning to be back in them. When Bill started photographing apples and orchards, he found he couldn't stop. When I began exploring the breadth and depth of the apple and its various stories and roles, the same fate befell me.

Imagine slicing open an apple – or even better, put this book down and go fetch a couple of apples and a sharp knife. Slice one apple from top to bottom to reveal its centre. The shape of the seed cavity revealed in this way reminds some particularly imaginative commentators of female genitalia, which turns the apple into a symbol of sexual desire. Slice the other apple horizontally, and in the heart of the flesh you'll see five seed chambers in the shape of a pentagram, one of the most widely used religious symbols in the world. Recently it's become most strongly associated with the occult, witchcraft and devil worship, but it has a long history as a doorway to the secrets of both good and evil. In early Christianity, it represented the five senses, or Jesus's five wounds on the Cross. Upside down, with two points facing up, it's a symbol of evil, the goat of black magic. The ancient Greeks saw it as representing the five elements of fire, earth, metal, water and wood, and today it's the primary symbol of Wicca in the same way that the Cross represents Christianity. That's an awful lot of symbolism in return for two simple knife cuts.

Between those five seed chambers you'd probably be able to pick out around twenty seeds. If you planted each seed in a separate pot or bed, only a few would flourish. But as a thought experiment, let's imagine you could give each seed the perfect composition of soil, light and water it needed to grow. This would be more difficult than you might think, because the perfect conditions would be different for each one. But if you succeeded in somehow nurturing every one of those trees to maturity, you'd have twenty quite different apple trees. If your individually tailored combinations of soil, light and water could make each tree happy enough to bear fruit, few of them would be pleasant to eat, and the chances of any of them being like the apple you sliced open would be minimal. Each one would be an entirely new variety of apple, with different preferences, needs and characteristics. Maybe, if you kept repeating the process for years, or even decades, you might find the next Golden Delicious or Kingston Black.

Like any living species, the apple's primary motivation is to survive and pass on its genes to the next generation. To do so, it's become a cunning seducer. Its beautiful blossom attracts bees and other pollinators to bring pollen from other trees to fertilize it. The most basic fruit or grain consists of a seed or embryo, and food to sustain it while it grows its own roots and shoots. Early apples were little more than this – tiny, bitter and probably toxic. But the apple learned some smart tricks over the years.

If the seed were to fall too close to the parent tree, parent and offspring would all be competing for the same nutrients and sunlight. Until recently, orchardists wanting to keep their trees in peak condition would plant them eight or nine feet apart. So in the wild, the apple learned to grow fruit that's shiny and alluring, begging to be plucked. Recent studies in the neuroscience of the senses show that the redder something is, the sweeter we perceive its taste to be. Shades of green suggest citrus refreshment. Deep in its genetic coding, the apple knows this, and puts itself on display to passing creatures.

While some humans leave the core and seeds when eating an apple, most animals devour the entire fruit. The seed inside is smooth and tear-shaped, and passes through a mammalian digestive tract with minimum fuss. Hours later, yards or maybe even miles from the parent tree, the seeds find themselves deposited on the ground in a pile of fertile manure.

With its incredible genetic diversity, each apple seed then behaves like Goldilocks in the bears' cottage. This ground is too wet. This ground is too dry. This is too rocky. This slope faces the wrong way. Right here is perfect. The forest ten miles north might suit seeds one, five and thirteen, but they'd never stand a chance on the grasslands to the south, which are perfect for seeds two, eight and nine.

Five hundred miles east, or across the ocean to the west, seeds twelve and nineteen, doomed anywhere else on the planet, might just find that perfect hollow where the temperature is cool enough, the big hill provides the right shelter from the wind, and the slope stops just before the dip where a late frost would spell disaster.

Malus domestica, the domesticated apple, originated in the Tien Shan mountain range in Kazakhstan. Its origins were proposed by the brilliant but mostly forgotten Russian scientist Nikolai Vavilov in 1929, and have been confirmed by DNA analysis in the last twenty years. Teams of American scientists have visited the apple forests of the Tien Shan mountains and encountered astonishing genetic diversity there. They have calculated that the apple as it has been bred and domesticated around the world contains as little as 20 per cent of the genetic material in the Kazakh forests, and that the apple's wild ancestors may hold the key to its future in an age where breeding and genetic technology are becoming ever more sophisticated.

But the apple has used its combination of seductiveness and genetic diversity to spread from Kazakhstan, along the old spice roads to China one way and through Persia and into Europe in the other direction and, ultimately to every temperate climate (with particular help from humans when it comes to navigating oceans). Drive along a country road in western England, and during the summer you'll probably see an apple tree in the hedgerow every mile or so. Each one is most likely the result of someone throwing a gnawed core from the window of a moving car. Each one is a new variety with a lottery player's chance of becoming a global superstar. It'll probably be sour or astringent, or particularly susceptible to pests, or poor of yield. But you never know. The first Granny Smith was discovered in a garden belonging to Maria Ann Smith in Eastwood, New South Wales (now a suburb of Sydney), in 1868. The first Bramley came from a seed planted by a young girl, Mary Ann Brailsford, in her garden in Southwell, Nottinghamshire, in 1809. The Yarlington Mill, one of Somerset's finest cider apples, was found growing out of a crack in the wall of a watermill in Yarlington, North Cadbury, in Somerset. Whenever an apple variety has the word 'seedling' or 'pippin' in its name, it's because it was discovered in a similar way.

Those familiar fruits are the result of thousands of years of crossbreeding, random at first, then guided by human hands. This long process has given us a fruit with astonishing genetic complexity and a wily resilience that helps explain our special attachment to it.

Once we've won the lottery and discovered an apple that tastes great, crops well and is resistant to pests, bugs and blight, it takes on a life of its own. The Bramley and the Golden Delicious are famous around the world. But surely this begs a nagging question: if I'm going to get an entirely different variety of apple by planting the seeds from my Granny Smith apple, how come we have so many Granny Smith apples? They can't all come from the same tree.

This is where human intervention comes in. Like our human genes, the apple's genes are complex. We may have some physical resemblance to our parents, but each one of us is our own person – unique. The only method for producing identical humans would be cloning. While we've managed to do this with sheep and dogs in the last few years, we've yet to do it with humans. But we've been doing it with apples for over 2,000 years.

To propagate a desired tree variety, you simply take a cutting from the tree you want to propagate and graft it on to rootstock – usually a young sapling selected for the characteristics of its root system and what these confer on the growth of the tree. The genetic instructions that determine the characteristics of the fruit come from the grafted wood rather than the rootstock. If the graft takes, you have a clone of your superstar tree. Graft a hundred cuttings from a Granny Smith tree on to a hundred different rootstocks, and you'll have a hundred Granny Smith trees. Every single Granny Smith tree in the world is a cutting from a cutting from a cutting of who knows how many generations, all of which trace a direct line back to that garden that now sits in a Sydney suburb. And if that sounds like a triumph of modern scientific ingenuity, what makes it even more incredible is that the technique was well established by the time of Alexander the Great.

Knowledge of grafting techniques spread unevenly throughout history, and was lost and found at various times. So imagine a time and a place where the principles of grafting and propagation were unknown. Imagine all you could do was plant apples from seed and hope that this one was a good one. Most of the trees you planted would produce sour fruit, or fruit that became corrupted on the branch. Imagine you found one tree that bore shining, healthy fruit, year after year, and that fruit was sweet, satisfying and nutritious. We now know that this tree is the product of genetic diversity and random mutations, just like us. But imagine what it must have been like not to know that. Imagine planting the seeds from your perfect tree and getting new trees that bore no resemblance to it while the first tree carried on, year after year, bearing perfect fruit. Is it any surprise the sacred apple tree is such a common theme in fairy tale and myth?

Jacob and Wilhelm Grimm were German academics, linguists and cultural researchers who specialized in collecting and sifting through folk tales that had been handed down orally for generations across Europe. In Central and Eastern Europe, the brothers found plenty of stories revolving around golden apples. These would invariably grow on a tree in a royal palace and be revered by the king. There's a common meme, interpreted in seemingly endless different permutations, in which the apples are stolen and a long quest begins to get them back, usually involving the hero being eaten and regurgitated by wolves, pecked by eagles or otherwise being killed, and sometimes reincarnated, in a variety of creatively grisly ways. Many of these tales are convoluted and difficult to follow, and it's not surprising that they've been forgotten.* But the image of the revered tree, alone in a private courtyard or centre stage in an

* In the oral storytelling tradition, stories were designed to occupy a group for a whole evening around the fire, to be spun out as long as possible. Read something like *The Odyssey* today and its endless digressions and unnecessary detail are at odds with the focus of modern narrative, where unnecessary words must be deleted and every paragraph has to move the story on or be cut. Some of these old tales make little sense until you think about them in the context for which they were created.

enchanted garden, is so common that it must surely be another example of the seedling that randomly – almost magically – turned out to be so much more special than all the others.

Perhaps the most famous Grimm story is that of Snow White, where an apple forms the linchpin in the conflict between the innocent beauty of Snow White and the jealousy of the evil stepmother, the Queen, the symbol of temptation and desire:

> QUEEN: And since you've been so good to poor old Granny, I'll share a secret with you. This is no ordinary apple. It's a magic wishing apple.
> SNOW WHITE: A wishing apple?
> QUEEN: Yes. One bite, and all your dreams will come true. Now make a wish, and take a bite.

At first Snow White is hesitant, but when the Queen pushes her, slyly, she admits there *is* someone she loves, and wishes that 'he will carry me away to his castle where we will live happily ever after'. Having admitted love and desire, she bites the apple and falls into a deathly sleep.

As you might expect, the original story is darker and more bloodthirsty than the Hollywood adaptation. In the version set down by the Grimm Brothers, the huntsman charged with taking Snow White into the woods and killing her finds he cannot do so, and brings back the lungs and liver of a young boar to convince the Queen the girl is dead. Believing the entrails to belong to the young girl, the Queen eats them. When she discovers Snow White is still alive, she tries not once, but three times, to kill her: first with a silky, laced bodice that she ties too tightly, then with a poisoned comb, and finally with the poisoned apple. After the first two attempts, the dwarves manage to revive Snow White. On the third occasion, the magical apple is half white, half red. The Queen, disguised as a farmer's wife, cuts the apple in two and eats the safe, white half herself. Snow White takes one bite of the poisoned half and immediately falls into what is, effectively, a coma. This time, the dwarves – unable to find the cause of her ailment – cannot revive her, and place her in a glass coffin.

Possibly the most unintentionally disturbing part of the original story is when a passing prince who, in this version, has never met Snow White before, spies the seemingly dead woman in the glass coffin, takes a fancy to her and asks the dwarves if he can take her with him. The dwarves don't think there's anything weird about this and agree. The prince has his servants lift up the coffin, but as they carry it away they stumble on some tree roots. The jolt forces a partially digested piece of apple to become dislodged from Snow White's throat, and she wakes up. Instantly, the young couple decide they will marry. In the Disney version, the spell can be broken only by true love's kiss. A handsome prince who previously saw Snow White and heard her sing has fallen in love with her. His kiss revives her, and everyone lives happily ever after.*

Why has this one fairy tale become world famous while

* Apart from the wicked Queen of course. In the Disney film, she plunges to her death from a cliff while being chased by the dwarves. In the original tale, she is invited to the wedding of Snow White and the prince, where, in an act of terrible revenge, the happy couple force her to put on shoes made from burning iron and dance in front of the other guests until, in unbearable agony, she drops dead.

so many of the other Grimm fairy tales have been all but forgotten? I'll concede that the dwarves have something to do with it. But might it also be that the apple, the embodiment of temptation and desire, appeals to us? Because it's not just in Middle European fairy tales that such a symbolic relationship crops up. It's consistent across pretty much any mythology or religion in the world's temperate regions. I wanted to know why. But as well as delving into ancient myth, I wanted to learn more about the real world of apple cultivation, too. From the first day I did so, I found that mythology and horticulture enjoy a much closer relationship than I could have imagined.

2.
Beltane

A smudge of pale blue is just starting to edge the sky when the alarm goes off. I dress in a daze, on autopilot, and half an hour later I'm in the back of a car, barrelling down winding lanes, startling rabbits and still-groggy pheasants. I know how they feel.

The horizon is now silver. We're racing against the sun, or at least that's how it feels. But it's not a race: we know exactly where and at what time we'll meet the sun; we just have to make sure we aren't late. So long as we meet nothing larger than a pheasant coming the other way, we should be fine.

By the time we park, about halfway to the summit of May Hill, it's light enough to see clearly as we walk the rest of the way. Rabbit holes that would have proven treacherous half an hour ago are now clearly visible in our path, perforating the closely cropped grass. We enter the treeline and walk through bracken, wild garlic and bluebells, a violet cloud hovering a foot above the ground. Ponies watch us go by impassively, as if waiting for us to get out of earshot before carrying on their conversation. They needn't worry: soon the sound of screeching from the trees around us is so loud we can't hear ourselves speak.

I've always thought of the dawn chorus as a genteel song,

and this is supposedly the perfect time of year to hear it. Today is 1 May, a Friday, and since 1984 the first Sunday in May has been International Dawn Chorus Day, when wildlife groups organize events to listen to the song.

When the first tweets began, I imagined some kinship with the pipits, woodcock, and raven, which live on May Hill, a solidarity and shared purpose in greeting the returning sun. But if this is song, it's less of a chorus, more a feedback-drenched screech with the amps turned up to eleven, early Jesus and Mary Chain rather than classic Vaughan Williams. The birds sound as if they're warning each other that strangers are near, or even telling those strangers in their most colourful avian language to bugger off.*

As we clear the treeline and approach the summit, the view begins to creep around my field of vision, closing to form a 360-degree panorama. May Hill straddles the border between Gloucestershire and Herefordshire, forming part of a ridge that separates the Severn and Wye rivers. It's the tallest hill for miles, and on a clear day offers views of thirteen counties.

We can't quite see all thirteen in the silver-blue pre-dawn light, but we can see the three that matter to me this weekend. Gloucestershire, Herefordshire and Worcestershire

* Later, I discover that the meaning of the dawn chorus was never as romantic as I imagined. The pre-dawn hour is a time when it's light enough to see, but still too dim to hunt insects, which aren't flying yet anyway. So the birds spend their time shouting to claim or defend territory, or to attract mates – and this is the height of mating season. A ritual I imagined as some kind of hymn to the returning sun is in fact the avian equivalent of the preening behaviour you'd see on a Saturday night in a Nottingham nightclub.

often refer to themselves simply as 'the Three Counties', and together make up one of Britain's richest horticultural areas. Along with Kent, it's one of the two key hop and apple growing regions in the UK – the two crops prefer similar climates, and are both close to my heart because they each produce one of my favourite drinks – but there's a huge eating apple crop grown here too. There's a Royal Three Counties Agricultural Show with its own show-ground near Malvern, where Bill Bradshaw and I have judged the Three Counties Cider Competition in the past. Until a few years ago, I'd never given much thought to what lay between Birmingham and the Welsh border, but the more time I spend here, the more I fall in love with its green slopes, narrow lanes, high hedgerows and quirky pubs.

May Hill's summit is a broad dome shape, topped by an isolated clump of tall trees. This makes for a striking sight and, like any curiously shaped natural feature, the hill gathers stories around it. It stands over the three counties, protecting their many orchards, and the local myth is that you can only make decent perry – cider's pear-shaped cousin – if you're in sight of May Hill. It's surprising how often you can look up from local orchards and find it watching. The Poet Laureate John Masefield first found fame with his long poem *The Everlasting Mercy*, which he wrote in 1911. Born just up the road from here in Ledbury, he saw the trees on May Hill as a ploughman and his team, and used it in one of the poem's most important passages:

> Near Bullen Bank, on Gloucester Road,
> Thy everlasting mercy showed

The ploughman patient on the hill
Forever there, forever still,
Ploughing the hill with steady yoke
Of pine-trees lightning-struck and broke.

I've marked the May Hill ploughman stay
There on his hill, day after day
Driving his team against the sky,
While men and women live and die.

Almost fifty years after Masefield's death, the trees are still there but, on the drive over, to me they resemble a gelled quiff. As we finally near them, I see that this morning they're accommodating around twenty tents, offering protection from a chill wind that becomes a quiet roar in the branches.

On the other side of the copse, facing east, around 200 people are gathered, huddled against the cold. Some are wearing garlands of flowers in their hair, which contrast oddly with utilitarian coats, scarves and gloves.

The sky is patchy but reasonably clear, and a fat band of gold and red starts shooting shafts of light into the cloud. 'It looks like we might be lucky this year, Hugh!' says Norman Stanier, clapping his son on the back.

Norman is somewhere in his sixties but looks much younger, thanks largely to the smile that occasionally leaves his mouth but never his eyes. He has the kind of cheerful optimism that might eventually get irritating if the person expressing it wasn't Norman.

'I've been up here about seven times now and I've only seen the sun once,' Hugh tells me. 'Last year the rain was horizontal. There were some environmental protestors who came here as some kind of holiday, under the trees in bivvy bags. Everyone felt sorry for them and gave them tea and soup.'

I wish someone would give me tea and soup. It's insanely cold.

Talk of food reminds Norman of his most important task. Later this morning, breakfast will be served in Putley Village Hall, just down the road from the orchard where Norman lives, and where we've just risen from a savagely curtailed night's sleep. Admission to breakfast is by ticket only, and the tickets are reserved for the brave souls who are here to greet the dawn. That means they weren't on sale beforehand. You can buy them only if you're up here on May Hill early enough, from the Breakfast Man.

'Is this him?' asks Hugh.

A tall, grey man wearing a big hat and a jacket made of multicoloured rags is walking towards us, carrying a sheaf of leaflets.

'Are you the Breakfast Man?' asks Norman, like a spy in search of a password.

'No, I'm handing out details of our programme of events over the summer,' says the man from the Forest of Dean morris dancers. 'I know the Breakfast Man is up here somewhere, though. I think he's wearing a smock and his eyes are a bit close together. His name's Rob.'

Norman nods sagely, and goes off in search of Rob, the narrow-eyed smock-wearer. Desperate for anything hot in the meantime, I spot a single table erected beneath the trees that looks as if it might have refreshments for sale. It turns out not to offer tea or coffee, but beer.

'Do you want one?' asks the man from Hillside Brewery.

'Bit early for me!' I reply.

'It's free,' says the brewer.

And so, at 5.30 a.m. on the morning of 1 May, I discover the crisp, clean and surprisingly warming virtues of India Pale Ale as a breakfast drink.

'He's sold out already,' says Norman, returning from his quest.

'There's beer,' I offer. 'It's free.'

Norman declines. Well, he is driving.

I first met Norman Stanier three years ago, when I was researching my last book. Norman owns Dragon Orchard on Marcle Ridge in the tiny village of Putley, and most of his apples are used to make multi-award-winning cider. When I arranged a return visit to come and learn more about apples and the orchards they grow in, rather than the drink they make, I had no idea this would involve drinking beer this early in the morning, flirting with hypothermia on a hilltop.

The light now has a luminous liquidity to it, and if I had a few more of these excellent beers, I could probably imagine it dancing invisibly with the three morris sides taking it in turns to perform in front of the shivering crowd.

Morris dancers are another feature of English life with which I've become more familiar since I've started spending time around apples and orchards. For many people, especially city dwellers, morris dancers are faintly embarrassing figures. I've always found it curious that so many countries have their own unique traditional dress, folk music and dance, and that we English resist the idea. Go to an Indian restaurant, get in the back of a Turkish driver's minicab, or even go to a proper Irish pub, and you'll hear music specific to their culture. See it live, and it will probably be performed by people wearing traditional dress. The English are unique in being uncomfortable about our own version, to the point of trying to pretend it doesn't exist. On international stages such as Eurovision, England is typically represented by a bowler-hatted and pinstriped city gent rather than someone in traditional dress like most other countries. The clothes and hats of morris dancers do look a bit silly. The dancers wave white handkerchiefs at each other, a celebration of limp-wristed surrender. They *skip*. For a nation that's always had trouble getting in touch with its feelings, this is not only un-English, but unmanly. Far better to say to the rest of the world that we're best typified as vicious capitalists rather than admit to any kind of folk history.

Over the past few years, I've done my best to be more objective about morris dancing, to try to look beyond my urban prejudices. In 2008, I accidentally found myself in Hastings on May Day Bank Holiday Monday and witnessed a mass procession of morris sides (I've even learned the correct terminology) from across the country. Some of them weren't awful. In fact, some – particularly the all-female sides – were actually rather impressive. The hankies had been replaced by big sticks, and they didn't just tap each other's sticks gently, they put all their weight behind them, splitting the air with whacks and cracks and throaty, guttural screams. It was as if they were channelling energy directly from the earth, releasing it into the sky like lightning rods in reverse. This really doesn't have to be shit, I thought to myself, and during the times I've watched morris dancing since then, I've tried to understand it from that perspective.

But it's not really working for me this morning. The sticks are out rather than the hankies, but it's all quite genteel and decorous. As the latest dance ends, a leading morris man welcomes us to the hillside and asks us to wave at our friends on other hills in other counties across the valley. We can't see them, but he assures us they're there. Then, just as the first gleam of the sun bleeds over the horizon, he turns his back to it and leads off on another dance.

The sun creeps up at 5.38 a.m., and it's picture-perfect. The sky is clear apart from a few clouds towards the horizon, and these erupt in salmon pink. Beams of light shoot upwards, just as they do when the sun is depicted in a cartoon. The rest of the sky glows in shades of violet, sapphire and gold, and I forget the cold, forget everything, forget who and where I am.

And no one else seems to notice.

I was expecting some kind of ceremony, with people facing the horizon and spreading their arms to embrace the sun, maybe even chanting something. But no. The morris men have started a new dance. They seem to think we've got up in the middle of the fucking night and walked up a great big fucking hill in the freezing fucking cold just in order to watch them. They're actually standing between the crowd and the horizon, and I have to walk around them, past them and down the slope a short way, to see what we supposedly came to see. Typical bloody morris: trying to upstage the sun.

But as the golden light hits my face, I feel elation and relief. I've attempted to stand a little outside this, the visitor from the big smoke observing quaint country customs, but the emotions I'm feeling are real, as real as those of someone watching a similar sunrise at Stonehenge 4,500 years ago. We know the exact time at which the sun will rise on any given day, and I find that tremendously reassuring. It imposes order on the world – in an old and deep and incontrovertible way. As the sun climbs and grows stronger, so does my joy. Summer is here. The leaves on the trees around me are only weeks or days old, still fresh and tentative, pale green and bright. The land is coming back to life. The certainty, the very predictability of it, is the source of its wonder.

Down in the orchards around May Hill, this is what the last few weeks have been leading up to. Through the winter, the apple trees sleep, and never has the phrase 'stark naked' been more appropriate than to describe the bony, wizened skeleton of a leafless tree in full slumber. As the temperature starts to rise and the days grow longer, the buds grow, until four or five small, tight blossoms appear in a cluster around one big central bud. As the fledgling blossoms start to emerge, the leaf buds spring into action, showing small tips of growth known delightfully as 'mouse ear'. A week or two later, the apple blossom has transformed the valleys of the Three Counties into a sight so beautiful it could melt the cold, dead heart of a Foxtons' estate agent.

A Druid wearing a green felt cloak, jeans and trainers somehow jumps into a brief gap in the morris dancing. 'I should know this off by heart by now,' he says, unfolding a creased sheet of paper and reading out a speech welcoming us to the festival of Beltane.

In the Celtic, pre-Christian tradition, Beltane stands opposite Samhain, the old version of Halloween. These two festivals divide the year, each a month and a half after the equinox. Samhain marks the end of the harvest season and the start of the 'dark half' of the year, and honours death and endings. Six months either side, Beltane celebrates beginnings and the return of the light.

Beltane is one of the few pagan festivals that Christianity didn't attempt to subvert and incorporate, which means it still retains a hint of danger. The most famous representation of it in modern times is in the film *The Wicker Man* (1973), which culminates in Edward Woodward's devoutly Christian cop being burned alive inside a giant wooden effigy. The clash in the film between ancient and modern, pagan and Christian, has resulted in the phrase 'It's all a bit Wicker Man' entering the vernacular to describe anything unusual or weird about the countryside, a rough equivalent to imitating the *ding-aling-ding-ding-ding-ding-ding-ding* duelling banjos from *Deliverance* to imply inbred, redneck deviancy.

There have been fire festivals around the start of May since the beginning of recorded history. Like the festivals of light of which Christmas is a part, wherever humans experience the cycle of the seasons – what the pagan tradition describes as the Wheel of the Year – we come up, independently or influenced by others, with similar ideas for celebrating important parts of it.

Pagans often expressed this with reference to a god and goddess, sometimes the Oak King and the Holly King, and more recently the May King and Queen. The May Pole – originally a fairly obvious signifier of fertility – has in later years been sanitized, and the May Queen has been reinvented as pure and chaste. But go to any May Day celebration and the symbols of the old festival of renewal, the return of light, the rising sap and the awakening of the fertile land, are still there somewhere. Despite (or perhaps because of) little media coverage and no obvious commercial sponsorship, festivals such as Jack in the Green in Hastings attract tens of thousands every May Day Bank Holiday, with people painting their faces green and threading ivy through their hair, marching the 'Jack' – a twelve-foot-tall green man representing the May King – up the hill to the castle before sacrificing him to release the spirit of summer.

The thread of this tradition that is specifically preserved as Beltane often begins the night before the first of May, and usually involves the lighting of fires built from the wood of nine sacred trees, each adding a specific power to the ritual, such as wisdom, joy or purification. Apple wood is added to celebrate love, which seems fitting here in the Three Counties, where the blossom urges thoughts of romance and weddings. This is the time when, as well as celebrating the trees coming back to life, the Celts drove their cattle between the fires to imbue them with life and fertility for the coming year as they took them back up into their summer pastures. In some versions of Beltane, people would then leap over the fires to get a little luck for themselves.

There's a tension in celebrations of Beltane when the desire to celebrate a lusty pagan festival meets a more everyday sense of English reserve and decorum, just as there is in morris dancing. In his masterpiece *Wildwood: A*

Journey Through Trees (2007), the late Roger Deakin describes how, while exploring the forests near May Hill, he learned of the Beltane celebrations carried out by the Cheltenham Pagan Gourmet Witches, a coven of eight or nine ladies who describe themselves as 'lardy women of a certain age'. They celebrate ancient seasonal festivals by taking large picnics and fine wines out into the countryside. On Beltane, their version of leaping over the cleansing fires is to step daintily over a small fire in a cake tin.

We don't even have that on May Hill this morning. It's National Trust property and they don't allow fires at all. If we did have one, I suspect we'd be huddling around and holding out our hands towards it rather than leaping over it.

The Druid reads a poem that begins 'Behold the waking sun of summer', but my hands are too cold to keep noting down his words. Later I try and check this line on Google and fail to get a single hit, so I'm guessing it's his original work. There's a polite smattering of applause when he finishes, and then, as surely as the sun rises, the dancers start again. The sun is high in the sky now, still unremarked upon by most of us here.

I wonder how many others on May Hill felt the delight I did in the rising of the sun, whether anyone has affirmed their hopes for the coming summer, whether Norman Stanier has asked the May King and Queen for a fruitful season. I've never celebrated Beltane before, but it's not the first time I've seen modern orchardists turn to ancient traditions and rituals to mark the turning of the wheel of the year and semi-seriously seek a bit of extra good fortune for their crop.

We're all too reserved to share such thoughts and feelings, but we've come together and had a bit of a laugh. In the codes of acceptable communication, everyone agrees it 'made a nice change'. Having failed to get to the Breakfast Man in time, our party departs May Hill, surprises a butcher just as he's opening up in a nearby village and retires for an epic fried breakfast.

3.

Dragon Orchard

The village of Putley doesn't quite convince you it's a village at all the first time you see it. It's one of those places that, if you're navigating your way to it by car, you say to the driver, 'I think we're nearly there now . . . We must be on the outskirts . . . Actually, I think we just went through it.' Putley has long been famous for growing fruit and, apart from its modest size, part of the difficulty in noticing it is that the houses don't really succeed in asserting themselves as a village at all but give up and accept that they're just random buildings scattered among the orchards, trees and hedgerows.

Herefordshire has always been good farming country. Here, five miles west of Ledbury, seventeenth-century farmers figured out that the microclimate created perfect conditions for growing fruit. The Malvern Hills rising in the east provide shelter from cold winds, as do the Bromyard Downs to the north. The Black Mountains across the border in Wales to the west create a rain shadow that blocks Atlantic weather fronts and gives the area an average rainfall of 600mm, half of that falling in the summer, which is the perfect amount for fruit trees. Closer to home, the tree-lined Marcle Ridge provides additional wind cover, and the south-facing slopes have deep, fertile soil.

Terroir may be a French word, but the concept of 'land taste' is just as important for other fruits as it is for wine grapes.

In the nineteenth century, the fight-back that saved the English apple industry from extinction began in and around these orchards. Victorian England was undergoing a love affair with the apple, with dinner-table conversations about different varieties that had been carefully selected for their perfect balance of flavour and crispness. The growing, affluent middle class aspired to country estates where they could entertain in a manner where every single aspect of a meal was an indication of wealth and discernment. The end of punitive taxes on glass in the 1850s meant glasshouses were now affordable as part of these estates, and in them all types of exotic fruit were grown. Rather than the needs of ordinary people living in the newly urbanized cities, cut off from their heritage on the land, it was the preferences of fashionable country-house dinner parties that drove the choices of gardeners and nurserymen. Apples were chosen for their novelty and fine flavour, which was fascinating and admirable if you were lucky enough to be able to appreciate it, and contributed to Britain becoming the most discerning and knowledgeable apple-growing nation in the world. But considerations such as yield and pest resistance were ignored. Apple growing was a glorified hobby rather than a serious aspect of British food production.

Towards the end of the century, the less affluent, but still moneyed lower middle class aspired to re-creating the fashions of the elite in their more modest city homes, and turned to foreign imports for help. The British reluctance to modernize meant that home-grown eating apples were often small and blemished. They were sold at markets all jumbled together, with no indication of what the different varieties were. French, Canadian and especially 'Yankie' apples were being grown with a more scientific, business-focused attitude, and were both cheaper and better quality than the crop coming up the road from Kent. Imports from across the Atlantic quadrupled between 1875 and 1879, and gardening journals in the 1880s reported that 'go where you will, American apples are in the fruit shop windows . . . The unattractive crab-like produce of the home orchards driven out by the cherry bright, clear skins and good looking proportions of the red Baldwins and Northern Spy.'

British agriculture was in depression, and farmers were looking to fruit, and apples in particular, to save them. Prime Minister William Gladstone urged farmers to plant fruit instead of corn to claw back some of the £5 million that was spent on imported fruit and vegetables in 1879 alone.

Dr Robert Hogg, Vice-President of the Royal Horticultural Society, understood that apple growers needed organizing better if this was ever going to happen, and came to Herefordshire. The county was traditionally the home of cider fruit, but the new railways meant the county could now also deliver fresh edible fruit to the big cities. Hogg met the Woolhope Naturalists' Field Club, founded in 1851 in Hereford 'for the practical study, in all its branches, of the Natural History of Herefordshire and the districts immediately adjacent', and together they worked to categorize the apples grown here and boost the cultivation of the best varieties.

At the time, most of the orchards in Putley were part of the Putley Court estate. In the 1870s this estate was acquired by Squire John Riley, a member of the Royal Horticultural Society and a passionate supporter of Hogg's project. Riley had travelled around the United States and observed how the more business-oriented approach of the Americans was enabling them to grow better apples than British orchards, ship them across the Atlantic and still sell them more cheaply in London than inferior apples grown a few miles away.

Riley hired a man called Harry Taylor to be the overseer of the estate and instructed him to plant it with fruit trees. Taylor replanted all the orchards on the Putley Court estate, and became good friends with the Riley family. Cider and perry were still made in abundance, but fresh fruit now also started to make its way to Birmingham and London by train. Norman Taylor, Harry Taylor's son, was the local stationmaster. He was also Norman Stanier's maternal grandfather. 'The railway was crucial for getting fruit to market quickly,' says Norman. 'Picked plums, for example, were graded overnight. We could have fruit in Paddington Station in four hours, and apples and plums could be on sale in London less than twenty-four hours after being picked. That's how it was for about a hundred years, until the Beeching cuts decimated the railways in the 1960s.'

After the First World War the Putley Court estate was broken up and sold. Thanks to his father's relationship with the squire, Norman Taylor was able to have his choice of the land his father had planted before it came on to the market, and built a house on the edge of what is now Dragon Orchard.

After our post-Beltane breakfast and a luxurious nap, Norman takes me on a tour of the orchard. He grew up in the house his grandfather built at the edge of the trees, before building his own eco-house in the middle of the orchard with his wife Annie in the 1990s. He sold the old family house to Simon Day, a wine-maker who fell in love with Dragon Orchard when he discovered it while walking his dog. Simon ended up founding the company Once Upon A Tree with Norman to make cider using its apples.

I don't know if Dragon Orchard has changed anyone else's life in such a dramatic way, but it's not difficult to understand what Simon saw here when he first encountered the place. With ordered, spaced trees that alternate between brilliant blossom in spring and shining fruit in summer and autumn, orchards don't have to try too hard to be beautiful. But Dragon Orchard goes beyond the call of duty. Norman Stanier has been working in it since he was a child, and over the years he's gently guided its natural beauty into something even more elegant, as well as more productive.

Most of the orchard sits on a very gentle slope, which is important for the right drainage. The soil is fertile, deep and holds moisture well. This means it isn't ideal for tractors, so it's no good for arable farming, but it's great for hops and fruit. 'And the climate was perfect before it was quite so variable,' says Norman.

'Climate change' has replaced 'global warming' in the terminology we use now. In his 2007 book *Unspeak: Words Are Weapons*, Steven Poole examines the way language is subtly coded to manipulate us, describing this as a cynical move to make it sound less scary. But there is some truth in it. At the National Fruit Collection in Brogdale, Kent, they're closely monitoring climate change, and orchardists tend to agree that it's not so much a case of everything getting warmer per se, more that it's becoming highly unpredictable. Warming is observable in the apple industry – blossom now starts to appear in Kentish orchards three weeks earlier than it did in the 1960s – but the effect of a global increase in temperature is far more complicated than that once it starts interacting with complex weather systems, and the increasing randomization of temperature and rainfall is hitting orchardists much harder.

Different varieties of apples blossom and ripen at different times, and in Dorothy's Orchard, closest to the house, Norman has planted a mix of varieties that are choreographed for an even spread of fruit and blossom throughout the year. He's also built a semicircular wall that shelters a long, curved bench from the wind and captures the sun. From here, the trees radiate out in their different shades of white and peach. At the end of each row, Norman planted greengage and quince, for no other reason than that they provide a pleasing visual coda. 'This is a celebrating space,' he says. 'When the fruit is ripening, the quince light up like lanterns catching the late sun.'

From Dorothy's Orchard, we walk up the slope, through alternating rows of Michelin and Dabinett apples. Most of this is cider fruit, planted in 1996 when the house was being built. The trees are about ten to twelve feet high and remarkably regular in size and shape. These are bush trees, planted on a specific rootstock that controls their height and vigour and makes them easier to harvest and prune. They also give the orchard a structured, stately elegance that's very different from ancient apple trees in the wild, or traditional orchards that go back a century or so. They stand close together in rows nine feet apart. The grass between each row is long and springy. Under the trees themselves, the earth is bare, but frosted with lichens.

The Dabinett blossom began a few weeks ago as tight buds of bright pink but is now brilliant white, while the Michelin is still stained pink on the outside of the bloom, with pale veins running through to the centre. Other varieties are still tight, almost scarlet, while still others have already been and gone. In some parts of the orchard, brilliant white petals fall steadily, like snow.

Apart from their aesthetic appeal, the more important advantage of bush trees is that they can be planted much closer together than traditional trees, dramatically increasing the yield per acre, which is vital for any commercial orchardist seeking to make a profit.

The riot of blossom in Dragon Orchard suggests that this year's crop will be a good one. The immediate sense of celebration it evokes draws people to admire orchards at this time wherever they grow around the world, and many apple-growing communities stage events in May to celebrate the glory of blossom. But these are also anxious times for the orchardist, and disaster could still strike from many directions.

Beautiful though Dragon Orchard is in blossom-time, the only thing that matters over these few weeks is that the blossom is pollinated. Biologically, that's the only reason it exists. Some plants can self-pollinate, but most apple varieties need to cross-pollinate, meaning they can be fertilized only by pollen from another variety from a different tree. The blooms attract insects that come to feed on the nectar inside. As the insects go from tree to tree, they carry pollen from one variety to the stamen of another.

The dangers are that an unseasonably warm winter can make blossom bloom early, before there are any insects flying, while a late spring frost could kill the bloom. Either can mean the insects won't come, so this is the tensest time of the year for watching the weather. If there's no pollination, there's no apple crop.

The most famous pollinators are bumblebees, which is why orchards often have beehives sitting in the corner. It's also the main reason why the recent rapid decline in the global bee population is such a concern for food security. If bees disappeared, apple trees wouldn't be the only plants to suffer – we'd be looking at a global food-supply catastrophe. Bees are not the only pollinators, but every orchard needs to offer an environment that's attractive to pollinators of some sort.

Also important is the management of the trees themselves, which is a hands-on job throughout the year. A growing tree produces both fruit and wood, using a finite amount of resources from the air and the soil. If you can restrict the amount of wood it grows, it has more energy to put into growing fruit.

That's why pruning a fruit tree is an art – there's far more to it than simply shaping the tree so that it's nice to look at or convenient to pick from. Cutting away at a tree stresses it and can make it feel threatened. So as well as pruning to direct more energy to the fruit, an unsentimental orchardist can also increase fruiting by making the tree fear for the survival of its line. A cider-maker in a traditional old orchard in Somerset once told me proudly that his trees never got enough water and were pruned severely so that they were continually stressed and unhappy – and that was the secret to his regular bumper crop. However, along the basalt cliffs of East Wenatchee, in the high deserts of Washington State in the Pacific Northwest of the United States, another apple grower proudly showed me his state-of-the-art irrigation and feeding system, and told me how careful he was to keep his trees happy and that was why his apples were superior.

There are merits to both arguments. 'It depends on what you're growing,' says Norman when I put this seeming contradiction to him. 'With cider fruit, you're paid by the tonne rather than how pretty or perfect the fruit is, whereas with dessert fruit, you want fewer apples but you want them to be bigger and prettier. Either way, you don't want to push the trees *too* hard.'

As well as directing the energy within the tree and controlling its stress levels, pruning shapes the tree to give it the best exposure to the elements it needs to grow well. 'My dad always used to say, "You're not growing apples, you're farming light and air," ' says Norman. 'Pruning and shaping the tree is all about maximizing the exposure the leaves and fruit get to the sun. It's like sleeping with the windows open rather than in a fug. You need the air to circulate, and light to power the tree. My grandfather always said that if you prune a tree right, you should be able to throw your hat through its canopy.'

Light aside, the tree also needs nutrients from the soil. Left to grow naturally, the topsoil is a churning mass of organic matter being broken down at the same time as plants seek it out to claim back the nutrients within it. Those nutrients are finite, so an orchardist wants to reduce the competition his trees face from weeds and other plants – hence the bare strips under each row of trees. But at the same time, you need a broader, more complex habitat to encourage pollinators. It's all a fine balance. 'When it comes to apple-tree cultivation, we act as if we know what we're doing, and that seems to work most of the time,' says Norman.

'Acting as if we know what we're doing' is a surprisingly apt description of how we work with nature. In recent years, we've made stunning advances in understanding how plants work and why, particularly in our understanding of plant genetics. But there's still a great deal we don't understand. For most of human history, we've relied on observation and experimentation: if we do *x*, then *y* usually happens, but not always. So we generate theories to explain why *y* usually happens, and more theories to explain what was different when it didn't. When we learn more about how things work, some of the earlier theories appear laughable. But take the example of the sun: if you have absolutely no idea why it sinks lower and the nights grow longer in winter, but you pray and make sacrifices and then honour the returning sun at festivals like Beltane, then, based on your minimal level of knowledge about how the solar system works, combined with an essentially logical brain, if you repeat the same actions again and again and always get successful results, it's not stupid to assume your actions had something to do with it. When you look at it from the perspective of our ancestors, the birth of the scientific method – carrying out actions and observing their results and then drawing conclusions about cause and effect – is little different from wearing your lucky pants on a date or to an important football match. Go back to a time when we had little or no store of scientific knowledge built up by generations of great minds, and the birth of early religions and, later, the development of scientific method are essentially two faces of the same spirit of enquiry.

Whenever we don't know why things happen, we develop hypotheses, and we do the best we can based on what we learn, or think we learn, from observation. Often, when we finally do understand what we're doing and why, it makes surprisingly little difference to our ability and success. As with anything in life, the good orchardist accepts and embraces the things they don't know. In this context, among all the spraying and pruning and clearing, climbing a hill at dawn to greet the returning summer sun feels like a sensible insurance policy. It links you back to the wheel of the year, and reminds you that you're part of something greater, spiritually and practically. And apart from that, when else are you going to climb to the top of the highest hill for miles around, and watch the sun come up and chase the mists from the fields and orchards of one of the most beautiful parts of the country, looking out as if you own the lot, at the very beginning of a new day and a new summer?

4.

The Rolling English Road

Just as Beltane and Samhain stand facing each other, looking back and forth across the seasons, so blossom-time and harvest measure and define the apple year. For over twenty years, a loose organization known as the Big Apple has celebrated both blossom and fruit in the cluster of small parishes dotted around the Marcle Ridge. There are seven in total, and while all have apples at their heart, the villages of Putley and Much Marcle are, in the words of Jackie Denman, 'the most appley'.

Jackie hatched the idea of the Big Apple after moving to Herefordshire from London in the 1980s. The year 1989 was British Food and Farming Year, 'Very commercial, sponsored by Weetabix, a big thing in Hyde Park', and Jackie wanted to create a celebration that local people could enjoy. 'I was living next to an old lady who had a very old orchard that was pulled out. It was too much hassle for her, and so it just disappeared. I thought that the only way to secure orchards was to find a value for them. Not necessarily a financial one, but a way of showing that people cared.'

This idea quickly took hold around the parishes. People who had worked in the orchards and retired or moved on began to remember what they had once found so special.

'They knew all about the old varieties. Some of them insisted they could identify an apple blindfolded! It all built word of mouth, and one idea led to another, with cider-making, apple exhibits and practical demonstrations. Norman's mum and dad were very involved. Generations have handed it on. That's the way things work around here.'

We sit talking in surprisingly warm sunshine outside Putley Village Hall, a new red-brick building in the heart of the village. Along with Putley's famous old church, it *is* the heart of the village. There's not even a pub here, although the new Parish Plan includes proposals for a pop-up pub inside the village hall on the last Friday of every month. (Norman is one of the drivers of that particular project.) Right now, the only refreshment is tea in Styrofoam cups, but as we sit chatting people arrive to set up stalls selling cider and apple juice, information stands about composting and wildlife habitats, and trailers selling burgers and sausages.

Jackie continues talking, quickly and efficiently. She's being very generous with her time, patient as I scribble my notes, but she's the heart of the operation and needs to be in at least four places other than the folding chair opposite mine. She doesn't seem fazed by the constant interruptions from people asking where to put this or where to find that. She combines the warm and open manner of a junior-school headmistress approaching retirement with the steely efficiency of a rock-concert promoter.

'It was just meant to be a one-off, but people started saying, "let's do it again". So in 1991 we got organized and drew up a constitution, deciding what we were about – just a small group of people from across the parishes. At first it was just Harvest-time, then we started Blossom-time in 1992. It's all about taking notice of the slightly odd ideas people come up with, and having people who are prepared to give them a go. Cider's an important part of it, but we're connecting the end product back to the place and the people. It comes from somewhere, and someone makes it.

'It's bigger than that now, though. We've had *Guardian* journalists down here in their pink boots, but we don't get a lot of exposure. It's not about that, it's organic. People feel connected to the village, and they come back year after year from places like Birmingham. I think over the last twenty years, people's urban lifestyles have become much more connected with growing things, with allotments, and so on. We're tapping into that, and we also feel like we're carrying forward something older, preserving things. One of our newest projects is recording an oral history of Putley and its apples. No offence,' she concludes, glancing at my notebook, 'but that's more important than writing things down.'

Saturday afternoon at Putley's Blossom-time weekend is not officially open to the public. It's a Bank Holiday weekend, and the main festivities run on Sunday and Monday. But Saturday is the part the locals are most excited about.

In 1787, the wardens of Putley Church wrote to *The Times*, boasting, 'Every man lives under his own apple tree and his own pear tree. When we meet, cyder contents us.' Now, the annual Putley Cider and Perry Trials are held in the village hall every Blossom-time weekend. If the cider from the previous year's harvest has been fermented traditionally, working slowly in barrels over the winter months, it should be ripe for drinking any time now. The event is not open to the public, but I'm allowed in as a special guest. Cider-makers are judged by their peers, who blind-taste each cider – just a sip, obviously – and score them. Originally intended for producers from the local area, the fame of the Trials has spread. There are 170 entries this year, including some brought swaddled deep in the luggage of people from Canada and Finland, who are hoping to be judged by and against their heroes.

Everyone tries to figure out which is their own entry, so they can score it highly. We're drinking from tiny, thimble-sized cups, but some of these people are professionals. They work quickly, and by the time I arrive the noise level is that of a party rather than a competition. A few of the faces around me are already flushed.

Because I'm here, and because I wrote a book about cider, the organizers ask me if I'll give out the certificates to the winners. I'm used to speaking on stage and it sounds simple enough, so I say yes. Local resident Liz Hurley is opening this weekend's chilli festival at nearby Eastnor Castle, so I'm in good company, even if I am the spectacularly low-rent option.

A few minutes before we're due to start, one of the organizers walks over to me and asks, 'All set?'

I nod and smile.

'So we were thinking, if you could just speak for about ten to fifteen minutes while we get all the certificates ready, that would be great.'

'Ten to fifteen minutes?'

'Yes, I mean we can't have you going on any longer than that. This lot have been drinking cider all afternoon and they don't really want to hear you. They just want to find out who all the winners are and then go home, so it's just long enough to fill the time while we get the certificates ready really. OK, maybe fifteen minutes, but no longer than twenty.'

I nod and smile again, a little less sincerely this time, then turn and run outside, forcing myself to stop rather than bolting down the lane. I find myself standing in the middle of a makeshift zoo, with ducks and baby rabbits around my feet. I pull out my phone. Thankfully, I have one precious bar of signal. I pray that Liz will pick up.

'I got asked if I'd give out the certificates to the winners of the cider competition and I said yes, and I'm on stage in five minutes and they just said they want me to speak for fifteen minutes and I don't have a speech, so please think of all the times I've been banging on to you about cider and apples and orchards,' I say as soon as she answers. 'All those hours. Is there anything that sticks in your head as remotely interesting or entertaining? Anything at all?'

'The stuff about the origins of the domesticated apple and how varieties spread is pretty amazing,' replies Liz.

'They've been drinking cider all afternoon.'

'Tell them the one about the Germans, and the one about Alex James out of Blur and the apple brandy,' she says quickly, hearing the panic in my voice. I hang up and scribble some notes.

Five minutes later, I'm on stage, telling the one about the Germans, and I get some polite laughter. I start to gear up for the one involving Alex James out of Blur and the apple brandy, the set-up of which involves me saying the word

'Somerset'. It gets a tirade of boos and jeers that a pantomime villain would give his hook for. I try it again.

'Somerset.'

'BOOOOOOOOOOOOOOOO!'

I've got them. I give a potted history of my cider book, explain what I'm doing with this new project about apples and orchards, and make sure to punctuate the whole thing with the word 'Somerset' every thirty seconds. It goes down a dream.

The awards presentation happens without further incident. The only moment there is anything close to drama is when one winner turns out to be fast asleep at the back of the hall. He's surprised and not best pleased to be prodded awake, and is still shaking his head in confusion as he takes his certificate from me and heads back to his chair.

'The full results will be posted on the noticeboard outside in about five minutes,' calls Jackie, as soon as I've handed over the last certificate. Immediately, the crowd heads out of the door and onto the grass. As soon as the last one is outside looking for the noticeboard, the door is locked and bolted behind them.

Inevitably, a local morris side is assembling in the road outside. Here, in the brightness of late afternoon, it's traditional, white-handkerchief morris, as much a part of spring in these parts as the apple blossom. There's some half-hearted tapping of sticks, but there's more energy in the hankies. Slowly, the magic of the day leaches away, and I finally realize specifically what it is about morris that bothers me.

Any public dancing has to fall into one of two categories:

it's either performance or participative; a spectacle worthy of an audience or a collective act. Usually, morris dancing is neither. There's little of the joy of spring in this, just a rather dull geometric progression. Too often, morris dancers demand that an audience stops what it's doing to watch them, and then rewards that audience by going through the motions, eyes downcast, faces devoid of expression, as if they're trying to remember what the next move is. They walk through the steps wearily, as if this is merely a rehearsal. There's no attempt to project. It's impossible to imagine ever paying to see this on a stage. I doubt anyone ever has.*

So if this is not dancing as *spectacle*, it should be dancing as *communion*. It doesn't matter if the dance is rubbish if we're all joining in, making fools of ourselves and laughing at the absurdity of it, as we would in a ceilidh or barn dance. But morris isn't communal either. Rarely is there any attempt to engage the audience, to share the ritual. And worst of all, the dancers look smug. And they have nothing to be smug about.

Having managed to get slightly woozy even on the small sips of cider I've been tasting, as the May sun begins to sink I decide to go for a walk to clear my head. There's no room for pavements on the narrow lanes around Putley, so I walk tightly beside sheer walls of hedgerow studded with dandelion and elderflower. There are paths across the

* The exception that proves the rule is modern English folk band Bellowhead, who, in the middle of their live concerts, have one bit where most of the musicians stop playing and do some morris steps instead, against a minimal but deliriously energetic backing. Imagine a heavy-metal guitar solo in the middle of a great song, only it's not a guitar solo, it's a traditional folk dance. The crowd goes *apeshit*.

fields and through the orchards, but you have to know where they are, and where they lead. There are no signs to tell you, so as a pedestrian rather than a walker you're always straining to listen for the long, drawn-out sigh of a car climbing a hill in the middle distance, the sudden flash of colour as it rounds a blind corner just in front of you.

I walk up the ridge, heading for Woolhope, three miles away, and the Crown, one of my favourite pubs in the country, which is holding a cider festival this weekend. As I climb, the valley stretches out behind me and it's full of blossom. Close up, row after row of neatly ordered trees, precisely spaced, run north to south, each as fresh and pretty as a prom queen. Further away, the blossom looks like fallen stars amid the green and then, further still, like an icing-sugar frosting on the landscape. I start to get an idea of the scale of fruit growing here. The diarist and gardener John Evelyn wrote in 1662 that 'all Hereford-shire is become, in a manner, but one intire Orchard', and that's how it seems this evening, with May Hill always visible, its Tintin quiff standing guard. And then I crest the hill and am on my way down into Woolhope, and there are no orchards to be seen. Instead, it's all fresh green fields dotted by sheep. The slope is steeper than it was on the other side. Maybe the soil is different. Whatever the reason, the change is dramatic.

As I weave my way down single-track lanes, I'm reminded of *The Longest Crawl* (2006), in which Ian Marchant travels Britain from south to north, via its pubs. The author lives just over the border in Wales, not far from here, and the book begins with him doing a walk not too different from the one I'm doing now. Marchant evokes

G. K. Chesterton's 'The Rolling English Road', arguing that 'this act of getting to and from the pub is central to an understanding of English topography, as well as English life'. It feels perverse to pull out my iPhone here in the stillness, as if I'm dropping psychic litter, but I want to read the poem now:

> Before the Roman came to Rye or out to Severn strode,
> The rolling English drunkard made the rolling English road.
> A reeling road, a rolling road, that rambles round the shire,
> And after him the parson ran, the sexton and the squire;
> A merry road, a mazy road, and such as we did tread
> The night we went to Birmingham by way of Beachy Head.

The third verse talks about flowers running behind him, 'and the hedges all strengthening in the sun', and I feel like I've won. I feel as if I could bound along these lanes, if the world is at my feet. I want to walk through this country all day. It's a feeling I get fleetingly from a great meal, or when a song turns and drops just so. Now, I identify it as delight.

'Male greenfinch in the top of that tree,' says our guide.

I look at some branches, seeing nothing that isn't part of

the tree itself. The guide hands me the binoculars he is carrying but hasn't yet needed, and points to a bare branch at the tree's crown. What I now know to be a male greenfinch stands proud, chest puffed out, chirruping like a fax machine. As I watch him, I feel hot breath on the back of my neck and turn to find a curious horse standing calmly behind me, hoping I might have some food.

This guided walk through Putley's fields and orchards is one of several that give the chance for townies like me to acquaint ourselves with the detail of the countryside. I'm desperate to explore the footpaths after my stroll along the roads last night, and I soon realize that people around here watch these hedgerows like I watch TV, read them as I read books. The hedgerows aren't just static – as the wheel of the year turns, different flowers bloom, fruit ripens, and species come and go. The locals walk regularly, keeping up with a soap opera by each pathway.

The hedges are mainly blackthorn, dense and prickly, studded with white blossom of their own. But they're mixed and merged with other plants, and I quickly learn to see more than the dandelion and elderflower I identified on yesterday's walk. I'm enraptured by bugle, a stunning blue wildflower that's in its element here on the first weekend of May, a multistoreyed bloom with petals piled on top of each other, a tower block for aphids. I start to ask the names of plants, and discover that I'm also a fan of borage, a light-blue, five-pointed star, and alkanet, similar but brighter, more vivid, with fatter, rounder petals.

Each of these is pointed out to me by Sally Webster, a passionate naturalist who finds as much excitement outside her cottage door in Putley as I would in an edgy corner of

East London. She's joined this guided walk even though she knows the area intimately, because there's always the chance of discovering something new. This is an exciting time of year for her, with the season on the turn and everything waking up. She's looking for specific things that will tell her everything is OK, and has been searching anxiously for a species of butterfly that was due to appear a couple of weeks ago. While they delight in their surroundings, everyone I talk to here is concerned that the seasons seem to be getting less stable, that the ordered pattern is fraying at the edges.

The benefits of paying closer attention to the hedgerows and margins are not just cosmetic. Up the road in Dragon Orchard, Norman Stanier has recently begun cutting the grass between his trees less often. He now mows every other row, then alternates the next time, so the broader ecosystem of the orchard can survive and prosper. He's also leaving wider margins at the edges, to re-create permanent natural habitats that attract pollinators to the blossom. 'It's about improving our relationship with nature,' he says, 'allowing it back into the orchard wherever we can.'

Norman is a deep lover of nature, as his orchard and his house demonstrate. So is Sally Webster. But their sentiments are increasingly shared across the fruit-growing industry and the more hard-headed, business-focused end of farming. As part of their joint Countryside Stewardship initiative, the Forestry Commission and Natural England now offer a 'Wild Pollinator and Farm Wildlife Package' which helps create year-round food, shelter and nesting places that wild pollinators need to survive. Evidence suggests that the right combination sown across just 3 per cent of farmed land brings significant benefits to wildlife.

Castle Fruit Farm in Newent, a few miles away from Putley, grows eating apples and plums for supermarkets. The constant price-pressure supermarkets place on their suppliers means efficiency has to be the primary concern, and the hedgerows around the orchard have always been trimmed back severely to maximize space. But now, orchardist Michael Bentley trims the hedgerows only one year in three. 'It's essential to provide nectar and pollen sources throughout the year,' he tells me when I visit later in the year. 'If you're planting new trees, you only get blossom on them in year two, so you need to provide continuity for the bees and other pollinators, to keep them here.'

When I was standing in that first orchard, back when Bill Bradshaw first infused me with his love of apples, I was struck by the interplay between man and nature, the co-dependence between them. After a weekend in Putley, I've learned that this relationship is a constant negotiation, ebbing back and forth. We've domesticated the apple in ways we haven't been able to do with other trees – as Colin Tudge points out in *The Secret Life of Trees* (2005), the squirrel has created a far more successful relationship with the oak tree than we've been able to, eating highly nutritious acorns when we cannot, and sometimes burying them and forgetting about them, thereby planting a new tree in a different spot. But we've been more successful with the apple, changing it and making it work for us. In return, we've propagated it across the world, taking it with us wherever we've gone. But we're still learning. There's still so much we don't know.

As the Beltane Festival showed, we don't yet know enough to completely discard prayer, ritual and ceremony. Even if our approach is tongue-in-cheek, the growth in popularity of such festivals over the last twenty years must have something more deep-rooted and serious driving it. Setting off to explore our relationship with the apple, the extent to which it permeates our myths and legends and its role in explaining how we became who we are, I find far more life in old beliefs than I was expecting.

5.

The First Orchard

When I was eleven or twelve, I made several attempts to read the Bible from cover to cover. This was partly because there were never enough books in the house, and partly because my dad, in a rare and half-hearted attempt to get me to go to Sunday school like a respectable child, promised me that the Bible contained some of the best battles ever fought, even better than those I read about in the *Warlord* and *Battle Action* comics he bought for me (and read first) every week.

I think I may have made it as far as the Great Flood once, but each time I tried, I quickly gave up. It wasn't just the endless begetting that sapped the soul; it was the strange syntax and random use *of* italics that made me wonder if I *understood* English *at* all. Anyway, I tried more than once, which means I've read the story of Creation in the Book of Genesis several times.

I thought I had a pretty good idea of what it told us: God created Earth in six days. On the last day he created Adam and, soon after, Eve – the first people. He placed them in the Garden of Eden, an earthly paradise where they enjoyed a period of blissful innocence with no cares in the world, until one day the devil, disguised as a snake, persuaded Eve to eat an apple from the tree of knowledge, which God had forbidden the couple to eat. Eve then persuaded Adam to have a bite. The apple gave them awareness and, I suppose, adult intelligence. They realized they were naked and covered themselves with fig leaves. When God saw them, he ejected them from Eden as punishment for their disobedience.

Thirty-five years later, I returned to Genesis and found the story to be not quite what I'd thought. I'd misremembered some parts completely. Searching for that fateful earliest apple, and with orchards and fruit trees at the front of my mind, I found both more and less than I expected.

On the first day, God created . . . well, the first day. He said, 'Let there be light,' and divided the light from the darkness, creating day and night and, therefore, the first day. On the second day he created Heaven, and on the third, dry land, grass, herbs, seeds, 'and the fruit tree yielding fruit after his kind, whose seed is in itself, upon the earth'. After creating plants, God created the sun, the moon and the stars, and then all the animals that fly, walk and swim (with whales the only creatures singled out for specific mention). He then caused 'a mist [that] went up from the earth, and watered the whole face of the ground'.

Having thus created the rain and the sun (and thereby the conditions for photosynthesis), God had established an environment in which plant life could survive. But he evidently felt that survival was not enough: he wanted his grass, herbs and fruit trees to thrive, and this meant careful stewardship, the interplay of nature and guidance. So on the sixth day of his labours, God created man from dust, because 'there was not a man to till the ground'.

God placed this man 'in a garden eastward in Eden', a

garden he himself had planted with 'every tree that is pleasant to the sight, and good for food'. What kind of flowers God planted, or whether this garden had beds, shrubs, grasses or lawn ornaments is not mentioned anywhere in Genesis, the emphasis is solely on the trees and their ability to provide fruit which is so nutritious that 'to you it shall be for meat'. Eden was not a 'garden' at all. It was an orchard.

In the middle of the orchard God planted not one but two special trees: 'the tree of life . . . and the tree of knowledge of good and evil'. God placed man – now known as Adam – in the orchard and gave him the specific instruction 'to dress it and to keep it'.

This is the point that seems to have been completely overlooked or forgotten over the history of Christianity. We imagine Eden as bliss, paradise, a place where work is unnecessary because everything man could want is at hand. But Genesis 2 clearly states that man was created by God specifically to look after the orchard He had planted. That was man's principal role on earth: to tend the trees that bear fruit. Adam had a job. The first man on earth was created specifically to be an orchardist.

Adam's reward for his labour was that 'of every tree in the garden thou mayest freely eat', with one important catch: 'But of the tree of the knowledge of good and evil, thou shalt not eat of it: for in the day that thou eatest thereof thou shalt surely die.'

Having established these simple ground rules, God added to Adam's labours. Having 'formed every beast of the field, and every fowl of the air', He brought them all to Adam 'to see what he would call them: and whatsoever Adam called every living creature, that was the name thereof'.

Now, God had created all these creatures in one day. But he was God. Adam, a mere man, was now tasked with naming every single one. At the last count, there are approximately 7.8 million species of animal on Earth, not including plants, fungi and the protozoa that Adam would not have been able to see (plus the animals such as unicorns and brontosaurus that never made it onto Noah's Ark, of course). In all the time since the dawn of humanity – whether you believe this to be less than 6,000 years by adding up the ages of all the people in the Bible or 1.8 million years by looking at incontrovertible archaeological fact – we've so far managed to name only about 954,000 of those 7.8 million animal species. This was an awful lot of naming for Adam to do – on top of looking after all the trees that are pleasant to the sight and bear fruit. Rather than pastoral bliss, Eden was starting to look like an outdoor workhouse.

God soon realized what an impossible labour he'd set Adam. So He decided to give him some help. Genesis 2:20–22 tells us:

> But for Adam there was not found an help meet for him.
>
> And the LORD God caused a deep sleep to fall upon Adam, and he slept: and he took one of his ribs, and closed up the flesh instead thereof;
>
> And the rib, which the LORD God had taken from man, made he a woman, and brought her unto the man.

God thus created woman to help man with his work. And it really is extraordinary how quickly things went wrong after this. While Genesis is incredibly specific about what happens when over the first six or seven days of Creation, the time period over which the rest of the action occurs in Eden is frustratingly vague. Some scholars have argued the whole thing takes place on one day, with Adam being created in the morning and expelled by teatime. Others suggest the action could have taken place over many years. Either way, it's fair to say we collectively imagine the first couple having a sustained period of prelapsarian bliss, enjoying the garden and all its fruits, naked and innocent before their fall. But the Bible story gives us no such scene. As soon as the person still known only as 'the woman' has been created, along comes the serpent to tempt her in the very next verses:

> 'Yea, hath God said: Ye shall not eat of any tree of the garden?'
>
> And the woman said unto the serpent: 'Of the fruit of the trees of the garden we may eat; but of the fruit of the tree which is in the midst of the garden, God hath said: Ye shall not eat of it, neither shall ye touch it, lest ye die.'
>
> And the serpent said unto the woman: 'Ye shall not surely die; for God doth know that in the day ye eat thereof, then your eyes shall be opened, and ye shall be as God, knowing good and evil'.

Tempted, the woman sees 'that the tree was to be desired to make one wise, she took of the fruit thereof, and did eat; and she gave also unto her husband with her, and he did eat'.

Their eyes opened, man and woman cover their nakedness with fig leaves and incur God's wrath. God tells

Adam that, from now on, life will be full of sorrow and sweat. Instead of tending the greatest orchard ever seen, he will now have to contend with dust, thorns and thistles, toiling to grow 'the herb of the field' to make bread. He curses woman with the pain of childbirth, telling her 'in sorrow thou shalt bring forth children; and thy desire shall be to thy husband, and he shall rule over thee'. This is when Adam – who is in charge of names, remember – calls his wife Eve, 'because she is the mother of all living'.

Paradise is over. God says, 'Behold, the man is become as one of us, to know good and evil: and now, he might stretch out his hand, and take also of the tree of life, and eat, and live forever.' He banishes mankind from Eden.

Armed with knowledge and insight, man is sure to attempt to eat the fruit that gives him eternal life. This, in God's eyes, is unthinkable: He places angels with flaming swords at the garden's perimeter, 'to keep the way of the tree of life', and to prevent Adam, Eve or any of their descendants from ever entering again.

The same story is also told across numerous places in the Qu'ran, although some of the details are different. Eden is Paradise in Heaven, set apart from the world. God creates Adam as his representative on Earth, and asks all the angels to bow down before him. Satan (known as Iblis) refuses to recognize Adam as his superior and is cast out of Heaven to live on Earth. In this version, God teaches Adam the names of all the animals rather than asking Adam to come up with them himself. Just as in the Bible, God creates Eve from Adam's rib, and places them both in Eden. But this Eden has only one forbidden tree: the tree of life. Satan/Iblis, in the guise of a serpent, persuades Adam and Eve together to disobey God and eat from the tree, each responsible for their own actions, promising them that it will lead the way to 'a kingdom that never decays' and arguing that Allah only forbade them to eat 'lest you should become angels or such beings as live forever'. When God finds out, He casts them out of the Garden to live upon Earth. But the difference is that, here, God reassures them that 'those who repent . . . and do what is right shall enter paradise', the seventh heaven, the beautiful garden.

Whether it was from the tree of life or a separate tree of knowledge of good and evil, Adam's apple sticks in the throat of all humanity like that little sliver of Kingston Black stuck in my own. We've been debating for over 2,000 years about what this story teaches us: is it about predestination or free will? Mark Twain loved it for what it told us about human nature, commenting wryly that 'Adam was but human – this explains it all. He did not want the apple for the apple's sake, he wanted it only because it was forbidden. The mistake was in not forbidding the serpent; then he would have eaten the serpent.'

This is the most famous apple of all, debated by scholars and philosophers, painted by the greatest artists of the Renaissance, copied and recycled in myths about temptation and eternal life throughout the centuries, in everything from Arthurian legend to Snow White. There's just one more problem with it: if Adam's apple truly is the most symbolic and fateful apple there has ever been, why does one of the most famous depictions of biblical scenes in the entire history of art – Michelangelo's ceiling of the Sistine Chapel – clearly and undeniably depict the tree of knowledge of good and evil as a fig tree?

PART TWO
Fruiting

May–June

‘Of trees there are some which are altogether wild, some more civilized.’

Pliny the Elder

6.

Survival of the Fittest

When the soft, white blossom petals have fallen, the fruit needs to 'set' on the branch. Petals gone, the heart of the blossom is brown and withered, but in the middle of each is a little green fruitlet, a few millimetres across, radiating spent fronds. If you were to dissect one, you'd be able to see the seeds starting to develop in its centre. This is a sign that the bloom has been successfully pollinated. If it hasn't, the tiny fruit will soon fall from the tree. This, then, is the first tentative indicator of whether the harvest will be good or not.

The exact time of petal fall and setting varies by a week or two depending on the apple cultivar.* While I'm at Dragon Orchard, most of the trees are still in full bloom, but a few are ahead of the game, and Norman shows me the fruitlets up close. The size and shape of the tiny apple fruitlets sitting behind the wreckage of the blossom remind me of rosehips, and the initially extraordinary fact that apples belong to the *Rosacae* family and are very closely related to roses starts to make more sense.

An apple tree takes no chances, and if the fruit does set, the tree will probably have far more fruitlets than we'd like

* 'Cultivar' is the term used most commonly to distinguish domesticated, named apple varieties from those that grow wild.

it to for our own purposes. A natural process known as thinning will see the tree shed a portion of the fruit it feels it doesn't need, but we step in and take it further. With cider fruit, the value is calculated by the tonne rather than how many or how big the individual apples are, so if the tree is happy, generally cider orchardists like Norman Stanier are too. But it's a different story if you're growing eating apples. We buy with our eyes and, even if that were not the case, we want to eat one, big satisfying apple rather than two or three that are no bigger than squash balls. (Many cider apple varieties are considerably smaller than eaters.) With eating and cooking apples, manual thinning, pushing the process further than the tree will go on its own, reduces the competition for light and food and allows the strong, prime fruit to grow bigger.

In that sense, thinning can seem a brutal process, killing the weak to help the strong. Author and journalist Frank Browning grew up on an apple farm in Kentucky, and in his book *Apples* (1998), he recalls the first summer his parents set him to work in the orchard. All the thinning was done carefully by hand, with people standing on ladders with their heads up in the branches of what were then mostly full-size trees. Most people new to the practice of thinning pulled off too few apples because they were fearful of ruining the crop. After Browning had spent an hour working his first tree, the young woman training him looked at the tree, looked at him and looked at the apples on the floor. 'What are you doing, feeling sorry for the little ones?' she asked. The stars of each bunch lay on the ground at his feet, the tiny, useless ones still on the branch.

Thinning is still one of the most labour-intensive tasks in

the orchard year, which makes it expensive. Cheaper options include using chemicals or strimmer-like machines that remove about 30 per cent of the blossom in May if it looks like it's going to be particularly heavy, but as yet there's no substitute for quality hand-thinning. At Brogdale Farm in Kent, home of the National Fruit Collection and the University of Reading's Farm Advisory Services Team (FAST), studies have suggested that an individual dessert fruit tree works at its optimum with precisely 120 fruit on the branch.

It's all about channelling the resources a tree has into producing fewer, bigger fruits, and it's important to get this right as early as possible in the season. An apple on the branch is 85 per cent water and 15 per cent cellular material. If you can increase the amount of cellular material by just 1 per cent, the apple is tastier, firmer and stores better. For the apple to have more solid material, the tree needs to have fewer apples to distribute its resources around. Cells divide in the fruit during only the first six weeks on the branch – after that the cells just get bigger, mainly by taking in more water. So in that vital six-week period the weather should, ideally be warm and bright, maximizing the energy the leaves can take in to feed the fruit, and the correct number of fruit needs to be fixed as quickly as possible.

As well as nudging the crop in the desired direction, thinning can reduce the biannual behaviour of an apple tree. Left to its own devices, a tree can produce a bumper crop one year, but this leaves it burnt out and sluggish the following year, so it takes time out. Thinning to regulate the number of fruit helps create a uniform harvest, year after year.

I'm surprised by how strongly fruit-growing relies on simple mathematic equations. If x is the total amount of energy a tree has, and y is the number of fruit on the branches, you can see how you arrive at the optimal number for y by figuring out how many you want to divide x by. But in spring, an orchardist is also trying to make x as big a number as possible, maximizing the amount of energy the tree has in the first place.

Most of the tree's energy comes through the leaves via photosynthesis, and this is why the leaves need to be protected from scab and mildew. 'Imagine having dirty solar panels – they're not going to get energy from the sun if they're covered. The leaves must be kept healthy,' Norman Stanier told me at Dragon Orchard. As well as scab and mildew, the tree also has to be protected from pests that might eat the fruit before we can, or even prevent it from growing at all. The constant battle with insects is so fierce that fruit entomology is a whole branch of scientific research in its own right.

The godmother of the discipline was Eleanor Anne Ormerod, the daughter of a noted historian who grew up on a large estate in Gloucestershire and from early childhood took an interest in the insects around her.* In 1868,

* Ormerod gained international fame for her work and received awards from the University of Moscow and the Société Nationale d'Acclimatation de France, as well as becoming the first woman ever to receive an honorary degree from the University of Edinburgh, where she was hailed as 'the protectress of agriculture and the fruits of the earth – a beneficent Demeter of the nineteenth century'. She was feted for her fearless dedication as much as for her patience and meticulous skills of observation. In a 1909 book about amphibians and reptiles, Hans Gadow recounts the experiment Ormerod carried out to explore the effects of coming into contact with the secretions from freshly caught newts. 'To

when the Royal Horticultural Society began creating a collection of insect pests, she made a significant contribution to it and, in 1877 she published a pamphlet, *Notes for Observations on Injurious Insects*, the first of a series of works which culminated in 1890 with *A Manual of Injurious Insects With Methods of Prevention and Remedy for Their Attacks to Food Crops, Forest Trees and Fruit*, a seminal text on the subject which reveals that, although knowledge and practice have changed, the problems facing orchardists haven't.

Professor Jerry Cross is the Science Programme Leader in the study of Pest and Pathogen Ecology for Sustainable Crop Management at the East Malling Research Station in Kent. He's the epitome of the engaging, slightly eccentric academic and, sitting in his book-lined office, I'm immediately thrown back to university, and the compounded arrogance and naivety that makes a student believe they can convincingly bullshit their way through a tutorial without having done the required reading.

Jerry introduces me to the works of Ormerod, almost successfully hiding his disappointment that I'm not already familiar with her when I'm supposedly writing a book about apples. He then takes me through the rest of the

personally test the effect, [Ormerod] pressed part of the back and tail of a live Crested Newt between the teeth,' wrote Gadow. He then quotes her write-up of the results: 'The first effect was a bitter astringent feeling in the mouth, with irritation of the upper part of the throat, numbing of the teeth more immediately holding the animal, and in about a minute from the first touch of the newt a strong flow of saliva. This was accompanied by much foam and violent spasmodic action, approaching convulsions, but entirely confined to the mouth itself. The experiment was immediately followed by a headache lasting for some hours, general discomfort of the system, and half an hour after by slight shivering fits.' Kind of puts my allergy into perspective.

great fruit entomologists: I nod and smile as he mentions Fred Theobald, Arthur Massey – good old Arthur! – Dicker and Cranham, and I realize fairly quickly, without him having to say it, that his name is the next on the list.

'We have new problems now, and there are some things we see differently,' he says. 'Take earwigs – a terrible plague on orchards, according to the Victorians. But now we try to encourage them. They don't do any primary damage to the fruit, just enlarge existing holes. And they're very important predators against other pests.'

With over 2,000 different species of insects, mites and bugs, an orchard is an incredibly diverse ecosystem. As scientists and orchardists have worked together to try to understand how to minimize the collective damage to the apple crop, they've learned to appreciate that damage can also be caused – or made worse – by interfering in one area without understanding the impact this will have on everything else.

Take permethrin, a synthetic chemical introduced in the late 1970s. It was very effective initially – in fact, too effective, because it killed all the natural predators in the orchard, which led in turn to a terrible outbreak of damaging spider mites that did more damage than the pests the spray had killed ever had.

Dichlorodiphenyltrichloroethane (DDT) was first synthesized in 1874, and it was found to be, in Jerry Cross's words, a 'wonderfully effective' insecticide by the Swiss chemist Paul Hermann Müller in 1939. It was then used in the second half of the Second World War to control malaria and typhus among civilians and troops. Its use as an insecticide spread after the war, and in 1948 Müller won the

Nobel Prize for Physiology and Medicine 'for his discovery of the high efficiency of DDT as a contact poison against several arthropods'. Norman Stanier remembers DDT being used in Putley's orchards. 'Men would wear sou'westers and waterproof coats and stand beneath the trees, spraying it upwards and getting covered in it,' he recalls, shaking his head.

In 1962, the American biologist Rachel Carson published *Silent Spring*, in which she catalogued the effects of indiscriminate DDT spraying on the environment and on human beings. She showed that it caused tumours in laboratory tests on animals, and that pesticides had unintended environmental consequences such as thinning of the eggshells of birds of prey. It stayed in the soil for years, and its intended targets simply built up resistance if it was over-used. Carson accused the chemical industry of spreading disinformation and public officials of accepting industry claims without questioning them. The book found support among the general public as well as the scientific community, and was hailed as a catalyst in the birth of the environmental movement. DDT was banned in the United States in 1972, and finally banned worldwide by the Stockholm Convention on Persistent Organic Pollutants in 2001. Just like the dawning realization of the importance of a full ecosystem for pollinators, the fruit industry is coming to terms with pest control and the need to work with nature rather than trying to dominate it.

Jerry Cross gives me a couple of wonderful examples of this. The orchardist's arch-enemy – Moriarty to the orchard's Sherlock – is the codling moth. Remember this old joke:

Q: What's worse than biting into an apple and finding a maggot?

A: Biting into an apple and finding half a maggot.

Well, the maggot in question is the caterpillar of the codling moth, which bores into the fruits of apple and pear trees during mid- to late summer. Eleanor Anne Ormerod discussed it in detail in her work in the late nineteenth century, and it's still a pest today. But entomologists like Jerry Cross have come up with ways of keeping those caterpillars out of the orchard without harming any other species of flora or fauna. During the codling moth's mating season, the female releases a pheromone that attracts the male. Entomologists have now isolated this pheromone and use it to confuse the male. It can be sprayed around the orchard at a steady rate so the male is bewildered, or put inside traps with sticky surfaces that trap the male: both methods mean the female is effectively stood up and never gets to lay her eggs.

But my favourite example is that of the aphid and the common black ant. These two species have a truly symbiotic relationship. Aphids suck sap from plants to get nitrogen. They get a lot of excess sugar, which they excrete as honeydew, an important food source for ants. So the ants protect the aphids from natural predators.

It's an incredible story, and I blurt out, 'That's amazing! So we want the ants and aphids in the orchard, right? So the ants keep harmful insects away?'

No. In fact, we don't want aphids anywhere near our apples. The weight of the honeydew they're shitting everywhere makes the leaves grow sticky and curl downwards. If the honeydew drips on to the fruit, it can cause sticky spots that promote the growth of black, sooty mould, and all that sucking out of nitrogen can mean the shoots of young trees that are heavily infested are stunted or malformed. But aphids are food for insects such as ladybirds and hoverfly larvae. One solution is to kill off the ants that protect the aphids from these predators (ladybirds may think they're a bit hard when up against aphids, but they'd get the crap kicked out of them by a black ant), but we're now trying to be cleverer about it. Jerry Cross has been working on a programme where ants are given all the sugar they can eat at the base of the tree, so they forget all about the aphids and their honeydew up in the branches and leave them to the mercy of the ladybirds and hoverfly larvae.

As I learn about pest control, it reminds me of international relations – fraught with unintended consequences and shifting alliances and deals, a place where the enemy of my enemy is my friend and has always been my friend and will always be my friend, until the next shift in the balance of power when it's suddenly the other way round. The whole thing is a delicate dance, one which we can't step out of once we've begun, the two partners moving one way and the other, bending and yielding, then pushing back. As soon as we assume we're the only dancer on the floor, things start going wrong. Whether we're trying to promote fewer, brighter apples, discourage pests or attract pollinators, we have to balance our immediate ambition against the broader ecosystem of which it's a part. We used to think we could control nature and establish complete dominion over it. But dumb old nature always fights back.

Orchardists don't like spraying – it's costly and it puts

chemicals into the ecosystem. But they have little choice, especially if they want to grow Britain's most popular apple varieties. 'It's a real problem here in Kent,' says Jerry. 'We grow lots of Gala and Cox because these are the apples the supermarkets want. But these particular varieties are highly susceptible to disease and fungus, especially if you get a wet spring, like we do here, so they have to be sprayed with fungicide fifteen to twenty times a year. You might ask, what about organic farmers – well, without chemicals, they spray even more! They're allowed to use natural substances like sulphur or copper, but they're not as efficient so they have to do it more often. The public thinks their organic fruit hasn't been sprayed at all, but it's been sprayed with huge amounts.'

A few years ago, I met an orchardist who came up with a different solution to the problem of pest control. Steve 'Bear' Bishop was a firefighter, battling forest fires in the Pacific North-west of the United States for thirty years. Sometimes it would take days to put out a blaze, and Bear and his colleagues would bed down in the hot ash, pulling silver, flameproof blankets around them while they tried to get a few hours' sleep. As he lay there, Bear dreamed of retiring to an orchard. When his dream came true, he took his expertise – and his blankets – with him.

Bear and his wife, Nancy, bought a plot of virgin rainforest in Port Townsend on the Olympic Peninsula, just across the water from Seattle. He cleared out the firs and cedars – not to mention the rocks – and used them to build a beautiful house.

Bear has a silver goatee and small, vividly bright eyes. He habitually wears a flat cap with the peak turned to the back, and a grey hoodie. He should look like a man dressing thirty years too young, but his energy and openness allow him to pull it off effortlessly. Nancy is never without a permanent wide, beaming smile, and together they give each other the energy and good humour of a couple their son's age. She's the analytical one; he's the adventurer.

'This was all silt and rock. I created this soil,' Bear told me with justifiable pride. Now, seventeen rows of apple and pear trees sit immaculately on trellises between soft, bouncy grass. Bear and Nancy planted the orchard in 2003 and got their first full harvest in 2008.

The orchard has been certified as fully organic, but Bear doesn't want to spray at all. Instead, he uses what he learned in his old job, where he saw how well trees bounce back from fire. He wraps each tree trunk in the same fireproof fabric he used to sleep under. Every other week, he gets in his miniature tractor and goes up and down each row with a flamethrower, scorching the earth beneath the trees, killing everything except the trees themselves.

Each row is irrigated with a black rubber water pipe, and precise irrigation helps keep the pests down once the apple-growing season gets under way. But this is America, and there are bigger pests than bugs and mites to contend with, such as voles, herons and rabbits. 'One time, I was trying to stop them so I brought my bow and arrow out,' says Bear. 'I don't like killing them, so I'd hit them with practice arrows. But they'd learn I wasn't serious about killing 'em, so they'd start hanging out by my irrigation pipe. I'm a pretty good shot, but I'd just shoot through the pipe. One summer I did that four times and Nancy took my bow and arrow away.'

The big drawback of our careful cultivation of the apple, of separating it from nature and trying to disrupt its relationship with natural predators, is that in trying to make it more hardy, we actually weaken it. A few popular apple varieties such as Golden Delicious, Gala and Fuji make up the majority of the world's apple crop. Different apples have their own particular advantages and disadvantages, and we happen to favour apples that are sweet and crisp but particularly susceptible to disease. Predators (or if we're honest, competitors trying to get the apple's goodness before we do) continue to adapt to the sprays and deterrents we use, building up resistance to them and becoming stronger, while the apple remains vulnerable. Some ecologists worry that, if we continue along this path, we could be heading for disaster.

And with that disquieting thought, my travels through Britain's orchards come to a temporary halt. The blossom has gone and the fruit is setting and the orchardists are spraying, but there's not actually that much for me to see. At key times of year orchards throw themselves open to people who come and gaze in awe. But as spring turns to summer, they're quiet and empty as the trees and the growers quietly get on with their work. So as I wait for the apples to ripen, I pass the time by returning to my desk, where I lose myself, completely and utterly, in the apple's history in both real and imagined worlds. It turns out to be as intriguing, complex and at times as beautiful a history as the orchards themselves.

7.

The Search for Eden

Leon Festinger never really intended to be a social psychologist. He was more interested in how environmental factors affected satisfaction with student housing at the Massachusetts Institute of Technology (MIT), but drifted almost by accident into a growing fascination with the behaviour of groups, and in particular how the way in which people communicate influences the decisions they make.

The experiment that made Festinger famous involved him infiltrating a Doomsday cult in 1954. A suburban housewife called Dorothy Martin claimed that she had received notification from an advanced alien civilization known as the Guardians from the planet Clarion that a flood was going to destroy Earth on 21 December of that year. Festinger and his colleagues joined her cult – known as the Seekers – and recorded the actions of its members, who had quit jobs and relationships and sold all their worldly possessions in order to win a place on a rescue spaceship the Guardians were sending to spare a chosen few. When Doomsday came and went, and both flood and spaceship failed to appear, the rational response would have been for the Seekers to turn on Dorothy Martin and accuse her of being a fraud. The proof was right there – she had given a specific date and details of the nature of our demise. She was clearly wrong. And yet what actually happened is that the devotion of the Seekers became even fiercer than before.

Festinger's celebrated book on the incident, *When Prophecy Fails* (1956), introduced the concept of 'cognitive dissonance' – our ability to simultaneously believe two thoughts that contradict each other, or take new information that contradicts previously held beliefs and work it into those beliefs. When the promised spaceship didn't appear, the Seekers became distressed. Seeking comfort from each other, they collectively decided that God had spared the planet, so the rescue mission had been called off. Such a close shave made the group all the more passionate about converting others to their cause. Festinger explained that the greater the number of people who agreed with them, the more their beliefs – which had been tested – would seem correct. Reinforcing each other as a close-knit group, their disappointment actually strengthened their convictions.

Of course, you'd never be that gullible. Would you?

Examples of cognitive dissonance surround us. Here's an example of which I was guilty before I started researching this book. I used to think the apple was a quintessentially English fruit, one that was native to these islands. And I also believed – I would go so far as to say that I *knew* – that the apple was the forbidden fruit in the Garden of Eden. How can both beliefs be true? Even if we don't believe the literal truth of Genesis, we accept that the storytellers who created it set the story somewhere in the Middle East, where the climate is very different from ours.

The Bible does not say that the forbidden fruit was an

apple at all. In the original Hebrew, the word used is *periy*, which translates literally as 'fruit from a branch'. When Adam and Eve became self-aware and covered their nakedness, they used the leaves of the fig tree, the only tree specifically named in the story. That's why Michelangelo made the leap and painted the tree of knowledge of good and evil as a fig tree. I'd always assumed it's because fig leaves are quite large, but it makes sense that Adam and Eve would have reached for the nearest leaves to hand once they'd taken that fateful bite.

Early rabbis variously argued that, as well as the fig, the forbidden fruit could also have been grapes – given the trouble they would later cause Noah – or the citron, a lemon-like fruit that symbolizes desire. The Jewish Book of Enoch, a religious tract that is not considered part of the biblical canon but still has religious significance, says, 'It was like a species of the Tamarind tree, bearing fruit which resembled grapes extremely fine; and its fragrance extended to a considerable distance.' Others speculated that it was probably a pomegranate. In the first few centuries of Christianity, it was never considered to be an apple.

The most common reaction among Westerners to finding out that the apple wasn't necessarily the forbidden fruit is to immediately switch sides: *of course* it couldn't have been an apple! Apples need a period of dormancy during winter; they need proper seasons, so they don't grow in hot climates like the Middle East. If we look at where the Garden of Eden is supposed to have been – somewhere around the Tigris and Euphrates rivers – apples can't grow there, but figs, grapes, tamarind and pomegranates can.

Apples were known in biblical lands – there are records

of them being eaten in ancient Egypt and recently discovered archaeological evidence of apple cultivation in Israel from around 1000 BC.* Outside the Book of Genesis, the phrase 'the apple of mine eye' is used frequently throughout the Old Testament to denote something of the highest possible worth, and the Song of Solomon's woozy 'Stay me with flagons, comfort me with apples, for I am sick of love' places the apple in the context of luxury and sensuality rather than sin. So could it still be a suspect?

To some extent, that depends on whether you regard the Bible as an allegory designed to help teach lessons about how to live, or a book of not only religious instruction but also real-world history and law, which is how it was mostly regarded until the Enlightenment. Before then, the story of Creation was the best account we had of how the world was formed and why we're here. Thinkers interpreted it in different ways.

If the Bible is allegorical, then we shouldn't worry too much about apples growing in hot climates. Genesis states that God planted Eden with 'every tree that is pleasant to the sight, and good for food'. In other words, because God is God and all-powerful, He can plant whatever trees He damn well wants. Eden defied the limitations of climate and *terroir*: if we're to take Genesis at its word, the garden must have contained vines, lingonberries, hops, bananas,

* While apples can't be cultivated as a rule in subtropical climates, they could have prospered on higher ground, where there is more variation in temperature. Even today, Israel's Golan Heights have bananas growing at the bottom of the hills, then avocados further up, then plums, with apples thriving at the top, where it's cool enough. Such places are, of course, the exceptions rather than the general climactic rule, but they do exist.

damsons, apples, oranges, mangos, cloudberries, and any other fruit you can think of, all growing side by side. If Eden can have angels with flaming swords guarding its approaches, then the correct temperature, rainfall and soil type for every different type of fruit on Earth is easy magic. And if we're going down this track, it's just a story anyway. To the authors, it didn't matter what the fruit was – it was merely a plot device.

The argument gets much more intriguing when you suspend disbelief and try to view the story as having really taken place. If the Bible is historical fact, dictated by God and transcribed by Moses, His most revered prophet, then Eden had a physical location on Earth. And if it did, we should be able to find those cherubim-guarded gates, even if the angels won't let us back in. The idea that Paradise is out there somewhere was, if nothing else, an obvious psychological response to hardship: a comfort around the campfire as the wind and the wolves howled outside the circle of light.

In fact, the Bible seems to give us some clues about where Eden might have been. For starters, it's in 'the East', which presumably means east of Egypt and Palestine. Then Genesis talks of four rivers flowing from it: the Pison, Gihon, Hiddekel (or Tigris) and Euphrates. We now know that the Tigris and Euphrates have their sources about fifty miles apart, in the Taurus Mountains of eastern Turkey, before flowing through Syria and Iraq to the Persian Gulf. Aside from a few geographical anomalies concerning the behaviour of rivers that weren't known to whoever wrote Genesis, we can conclude that the search area for Eden centres somewhere along the paths of the Tigris and Euphrates rivers.

If Eden was at the source of the two rivers, up in the Armenian highlands, the apple could have grown there. But because the rocky topography there doesn't sit well with images of Paradise, the consensus has been that Eden must have been somewhere further along the path of the two rivers, in the welcoming lands known as the Fertile Crescent, the birthplace of agriculture. Here, apples would have been one of the few fruits that would struggle to prosper, and so the fig or the pomegranate come into the frame again.

The problem with using reasoned argument such as this to locate Eden and identify the forbidden fruit is that logic doesn't always rub along well with devout religious belief. The theory that Eden might be hidden somewhere in the Tigris–Euphrates region was challenged by the inconvenient fact that this was also the location of humanity's first cities. It was one of the earliest regions on earth to be settled and mapped, and was notable for the total absence of angels with flaming swords. This is where cognitive dissonance really comes into its own.

French academic Jean Delumeau is the author of *History of Paradise* (1995), a haunting, breathtakingly thorough history of the quest to find the Garden of Eden. As I read it, I imagined a new Indiana Jones franchise. The search for Eden defied geographical reality and roamed across the whole world, with different fantastical theories competing to stay ahead of real-world discovery.

In keeping with the venerable tradition of literally taking as gospel the bits of the Bible that support your views while completely ignoring the parts that don't, if the great rivers specifically described in Genesis didn't lead us to

Eden, then Eden must be somewhere else and we must have interpreted the names of the rivers incorrectly. Simple! To escape the biblical flood, Eden must be atop a high mountain or on some magical plane halfway between earth and heaven, or even on the moon, even though the Bible – taken literally, as it was intended to be – doesn't say it is. One explorer definitely saw it in the Himalayas, but couldn't quite reach it. The only place it could possibly be if it was to preserve its inaccessibility from us was the North Pole – or possibly on some mysterious island. Over the centuries, such theories have come in and out of fashion, debated by the most brilliant minds of each age.

The great voyages of discovery of the fourteenth to the seventeenth centuries were in part motivated by a genuine desire to rediscover Eden. They did feed the flame of hope for a while, but as they mapped the world, they eventually snuffed out that hope altogether. Before the oceans were opened up, the common belief around the Mediterranean was that the flaming, hostile deserts to the south carried on indefinitely. Aristotle declared that the equatorial regions were uninhabitable. But some Christian scholars pointed smugly towards the blazing desert and said, 'See? *There* are your cherubim with flaming swords!' as if Paradise might lie beyond them. After the Portuguese invented the seagoing carrack and began exploring further and further down the West African coast, they discovered places that, rather than being uninhabitable, had a climate of perpetual spring, with days and nights of equal length, which started to sound promising.

Christopher Columbus sailed west, hoping to find Eden 'in the east' from the opposite direction. On his third voyage to the Americas, in August 1498, he believed he was close to success. Arriving in the New World, he confirmed that the region below the equator was full of lush vegetation, abundant rainfall and pleasant temperatures. It seemed obvious that they were nearing Eden, because this is how Genesis had described it and Genesis was true. When he arrived in Venezuela, the might of the Orinoco River convinced him that Genesis must have simply got the name of Eden's river wrong. He wrote to the Spanish King Ferdinand and Queen Isabella:

> I have never read or heard of so great a quantity of fresh water so coming into and near the salt. And the very mild climate also supports this view, and if it does not come from there, from Paradise, it seems to be still a greater marvel, for I do not believe that there is known in the world a river so great and so deep.

After one more voyage in search of his 'Paradise terrestral', Columbus died still believing that he was within touching distance of it, and the Portuguese continued to search in Brazil for the Garden of Eden for over a century.

They weren't the only ones hoping to find Eden in the New World. When they reached North Carolina, the English sea captains Philip Amadas and Arthur Barlow reported back to their commander, Sir Walter Raleigh: 'We found the people most gentle, loving and faithful, void of all guile . . . and such as lived after the manner of the Golden Age. The earth bringeth forth all things in abundance as in the first creation, without toil or labour.'

Inspired, Raleigh personally voyaged to Guyana. He

proclaimed that the Tropics boasted 'so many sorts of delicate fruites, ever bearing, and at all times beautified with blossoms and fruit, both greene and ripe, as it may of all other parts bee best compared to the Paradise of Eden'.

But if these explorations initially convinced Europeans that they were close to finding biblical Eden, eventually they destroyed the idea of its possible earthly presence altogether. As well as beautiful fruits and wonderful tame animals, there were also storms, earthquakes, diseases and other animals that were poisonous or dangerous in other ways, none of which were compatible with Eden. No part of the real world, no matter how beautiful, can ever match the images of Paradise we create in our imaginations.

In his *History of the World* (1614), Raleigh grudgingly conceded what all clever Christians and Jews were thinking but none wanted to say. Eden was no more:

> Although the garden itself cannot be found, inasmuch as the Flood and other accidents of history reduced the land of Eden to the state of ordinary fields and pastures, the place nonetheless remains what it was, and its rivers are unchanged.

Of course, the Garden had been real, but it was destroyed by the Great Flood. Some rejected such an explanation: God made Eden, and He would never allow his perfect creation to be destroyed in such a fashion. But by the seventeenth century, most who shared a painful nostalgia for Eden admitted that, even if it had ever existed in a real sense, it no longer did.

Eden moved all around the world – and beyond – in its successful attempts to avoid detection, through places where the apple could thrive and locations where it could not possibly grow.* The true identity of the forbidden fruit, when we consider all alternatives, is anyone's guess.

Whichever way I looked at it, I could neither prove the forbidden fruit was an apple nor prove that it couldn't have been. Even St Augustine of Hippo, early father of the Christian Church, tried to have his blessed cake and eat it by saying that the Bible should be read as both fact and allegory, so we really shouldn't worry too much about whether, for the purposes of the story, the fruit was supposedly an apple or not.

But this still doesn't explain why so many in the Western world are so certain that it was an apple, and have been for centuries. I needed to explore that aspect further, and it would take me from Christianity's Creation myth on a tour of myths and legends around the Mediterranean and up into northern Europe. But before I could say goodbye to Eden and embark upon that quest, God's orchard had one more twist in its story that would have profound consequences for the apple in the real world.

When we finally accepted that the Garden of Eden didn't exist on Earth, when there was nowhere left to search, we had to accept that there was no chance of ever returning to a Golden Age, or an earthly Paradise. The search for the Promised Land still continues in storytelling but takes on new faces, from C. S. Lewis's Narnia to Alex Garland's novel *The Beach* and the TV series *Battlestar Galactica*. But the acceptance that biblical Eden was no longer present on Earth marked the beginning of European

* The moon probably taking top prize in the latter group.

modernity. Having explored earth without success, thinkers and explorers turned their attention to trying to understand better what they *had* discovered. The bounty of the New World, they argued, meant that, for the first time since the Fall of Man and humanity's expulsion from Eden, the whole of God's creation could be analysed. Parts of it had been hidden from us until the great age of exploration, but now we could reunite the whole. If Eden no longer existed, we could re-create it ourselves, by bringing together all the scattered pieces of God's Creation and thereby gaining a better understanding of His work.

Early in the seventeenth century, first in Italy, then across Europe, gardeners began to create ornate and precisely ordered 'botanical gardens'. The general plan was to have a square garden consisting of four quarters, which represented the four corners of the earth, or perhaps the four continents of Europe, Asia, Africa and America. The beds were filled with as many different plants as possible, celebrating the diversity of God's creation, which had been scattered until now.

The first botanical garden in Britain was built in Oxford by the Earl of Danby in 1621. It featured 2,000 different plants, of which no more than 600 were native to Britain. John Evelyn, the prolific writer, diarist and gardener, whose work has, unfortunately been overshadowed by that of his more rakish contemporary, Samuel Pepys, took a close interest in botanic gardens, travelling to Italy to study them and then building one at his home in Wotton House, Surrey.

In 1664 Evelyn published *Sylva; or A Discourse of Forest-Trees, and the Propagation of Timber in His Majesties Dominions*, which he had delivered initially as a lecture to the newly formed Royal Society. The main thrust of the book was to categorize the main types of tree grown in Britain, and to urge more planting of these trees to provide wood for the Royal Navy. But appended to the main publication were two other works. One of these, the *Kalendarium Hortense*, is a guide to tasks that need to be carried out in the garden, broken down month by month – the first gardener's almanac. Imagine, if you can, an Alan Titchmarsh-type as some kind of religious teacher. Evelyn argued that our gardens should be made 'as near as we can contrive them' to resemble the Garden of Eden. Such gardens could be maintained only by human endeavour, and Evelyn was quick to point out what many had overlooked – even though God originally planted it, Eden itself relied on human beings to maintain it. He even suggests that our expulsion from Eden must have been the reason for its decline:

> As *Paradise* (though of *Gods* own Planting) had not been Paradise longer then the *Man* was put into it, to *dress it and to keep it*; so, nor will our *Gardens* (as neer as we can contrive them to the resemblance of that blessed Abode) remain long in their *perfection*, unless they are also continually *cultivated*.

'A Gard'ners work is never at an end,' he sighed, 'but this labour is full of tranquillity, and satisfaction' which 'contributes to piety and contemplation, experience, health and longevity'.

The botanical garden was the birth of modern plant science, a living laboratory driven by cataloguing, the analysis and collection of data, the observation of cause and effect, in as many different species as could be gathered in one place. But the motivation behind it was entirely religious – the re-creation of the Garden of Eden. Science and religion were two approaches to the same goal, belief systems that often stand opposed to each other today, but that can also overlap and work together: two aspects of the same desire to understand and make sense of the world. We may have lost Eden for ever, but Kew Gardens isn't at all bad.

John Evelyn loved the apple tree, but he didn't specify whether or not he believed it was the biblical tree of knowledge. However, it's almost certain that he did believe this, because in Britain and Western Europe it had long been depicted as such in art, literature and the poetry of Milton. Exploring some of these paintings and stories leads us further from Eden into a broader, older array of myths and stories in which the apple becomes a sort of 'everyfruit', standing as a symbol of all human hopes, fears and aspirations, to the extent that making sense of its role in Eden seems straightforward by comparison.

8.

The Real and Imaginary Apple

Reading Jean Delumeau's book about the search for Eden reminded me of a passage in *Generation X* by Douglas Coupland. In one of the book's most beautiful scenes, Andy, the main narrator, returns to his childhood home for Christmas. He rises early, while it's still dark, and fills the living room with candles of all shapes and sizes, everything he can find, from fat church pillar candles to little birthday-cake toppers. Finally, he allows his family into the room, where for just a moment, this group of bickering, dysfunctional people is 'dazzled by the beauty of the light', their eyes burning, 'if only momentarily, with the possibilities of existence in our time'.

I've seen Coupland read this passage several times at literary events. It's clearly one of his favourites, and his readers love it, too. He talks about how people approach him and say, 'I loved how you described that room. It was amazing, you captured every detail and it was exactly like my childhood home at Christmas.' At which he smiles and responds, 'I never said a single thing about the room. I didn't describe it at all. I only wrote about the candles. You filled in the rest of the description yourself.' It's a satisfying example of the intimate bond we form with the books we read, and the power of our imaginations to complete what's on the page, fill in the gaps and make the story our own by the simple act of reading it.

Historically, our relationship with Eden was no different from the one we have with Andy's parents' front room. Paradise is barely described in Genesis (2:8–10):

> And the LORD God planted a garden eastward in Eden; and there he put the man whom he had formed.
>
> And out of the ground made the LORD God to grow every tree that is pleasant to the sight, and good for food; the tree of life also in the midst of the garden, and the tree of knowledge of good and evil.

There then follows the naming of the four rivers and a description of the lands into which they flowed, and that's it.

Like some of Coupland's readers, Bible students filled in the space with their own details, almost unaware that they were doing so. Common elements of the Garden of Eden – none of them mentioned in the Bible – include pleasant warmth and perpetual spring, the best parts of all seasons in one, with no summer drought or winter ice. In this winterless garden, there is not only abundant fruit but also milk and honey – the only other foodstuffs alongside fruit that nature offers without the need for hard work. In the fourth century AD, St Ephraim the Syrian wrote of 'the light-filled dwellings, the fragrant springs . . . shadowy February smiles like May . . . December is . . . like August with its fruits, June like April'. Jean Delumeau's book is full of such descriptions down through the ages. But if this vivid detail isn't in Genesis, where did it come from?

The Bible wasn't the first religious book to describe the

perfect garden. The ancient Mesopotamian myth of Gilgamesh, which pre-dates the Bible by at least 1,500 years, contains a story with striking similarities to the events in Eden, complete with a sacred tree and a serpent causing trouble around it. So it seems much of the basic story was lifted from there.

As for the detail added by later readers, much of it strongly resembles the Happy Isles of Greek myth, much of which was also written down well before the books of the Old Testament. Greek tales often dwell on descriptions of a winterless paradise that became the final earthly abode for heroes and champions. In *The Odyssey*, Homer describes the gardens of Alcinous, King of the Phaeacians, on the mystical (and entirely fictitious) far-off island of Scheria:

> Here luxuriant trees are always in their prime,
> pomegranates and pears, and apples glowing red,
> succulent figs and olives swelling sleek and dark.
> And the yield of all these trees will never flag or die,
> neither in winter nor in summer, a harvest all year round.

The Greeks knew of no real-world gardens like this, but they were a popular and understandable fantasy at a time when work was hard and harvests were uncertain. Every time a mythical garden is described, the account lingers on the abundance of free food and drink avaliable, with no need for work. This is how Eden is fixed permanently in our minds and always has been – Paradise, a place of idle, innocent bliss – even though the Bible doesn't mention anything other than Adam's busy workload as orchardist and animal-namer. Sometimes deliberately, other times unwittingly, when ancient Bible scholars sought to describe Eden, they lifted their ideas from someone else's myth.

Unlike the ambiguity of Eden's forbidden fruit, in Greek myth the apple is clearly named, which surely helped its later migration on to Eden's tree of knowledge. Aphrodite, goddess of love and beauty, was often depicted holding an apple bough. The orb of dominion held by Zeus was an apple, and apple garlands were given as prizes for victory in games.

The golden apples of Greek myth are central to some of the greatest and most celebrated tales. Originally given by Gaia, otherwise known as Mother Earth, as a wedding present to Hera on her marriage to Zeus, they were planted in yet another mystical garden, this one in the far west, at the ends of the earth. These golden apples tasted of honey and were automatically replaced as soon as they were eaten. They could heal wounds, and if you threw them – which the ancient Greeks often did – they would always hit their target before returning to the thrower's hand. Like the fruit of every tree of life across all beliefs, they also conferred immortality on anyone who ate them. Such precious fruit had to be kept under close guard, which was a job given to the Hesperides, the nymph daughters of Atlas, the titan who held the weight of earth and the sky on his shoulders, and their pet hundred-headed dragon.

Despite being guarded so carefully, the apples were frequently taken from the garden. Whenever they were, tragedy ensued. Heracles (Hercules in Roman myth) was charged with stealing the golden apples from the Garden of the Hesperides as one of his twelve trials. The story of how he did so is a soap opera of epic proportions, which can only be appreciated when you remember that stories were often told orally and were designed to entertain people for hours on long winter nights with endless digressions, descriptions and embellishments. Reduced to its bare bones – Heracles succeeds in stealing the apples, entertaining the gods with his exploits, only for the goddess Athena to take them back from him and restore them to the garden – it's deeply unsatisfying. Like the God of Genesis who lied to Adam and gave him the free will that was bound to result in his fall from grace, the Greek gods dangled immortality (in the form of an apple) in the face of man before snatching it away, teaching humanity that its aspirations to be like the gods would always be futile.

But the most famous and catastrophic golden apple was the one that started the Trojan War. Eris, goddess of strife and discord, was annoyed at not being invited to the wedding of King Peleus to the beautiful sea goddess Thetis. Zeus snubbed her because – well – she was the goddess of strife and discord.

The king of the gods should have known better than to think strife and discord could be beaten so easily. In revenge, Eris inscribed one of the magical golden apples with the words 'to the fairest' and rolled it into the wedding feast. When the apple was spotted and its inscription read, the goddesses Hera, Athena and Aphrodite each argued that it was obviously meant for them.* The three goddesses asked Zeus to make a judgement on which of

* The 'Apple of Discord' remains in widespread use to refer to the crux of an argument from which greater consequences may ensue. In 1965, the publication of a tract named *Principia Discordia* marked the birth of a religion-cum-philosophy that venerates Eris as the goddess of chaos, rejects structure and order, and can't

them was most worthy of the claim. Faced with three very powerful women and a question that could have no possible right answer, Zeus did what any man, mortal or immortal, would do: he passed the buck to someone else.

Paris, a Trojan mortal, had previously impressed Zeus with his fairness and reason, so Zeus declared that Paris would make the ruling. The three goddesses approached Paris and undressed before him. Hera offered him dominion over all of Europe and Asia; Athena offered great wisdom and skill in war; and Aphrodite offered him the hand in marriage of the most beautiful woman in the world. Perhaps swayed by the pressures of the task immediately before him, Paris accepted Aphrodite's offer and declared her the fairest.

What Aphrodite had neglected to tell Paris was that Helen, his prize, was already married, to King Menelaus of Sparta. When Paris claimed his new trophy wife, sneaking Helen from her royal chambers and taking her home to the city of Troy, he sparked the beginning of the Trojan War, the longest and bloodiest conflict in Greek mythology – a conflict in which all three goddesses, victor and spurned, played an active role.

Pick an ancient mythological or storytelling tradition from anywhere around Europe and the Mediterranean, and the apple crops up as a mother lode of meaning, so much so that its different aspects start to contradict each other. Its roundness represents the female breast as well as the womb, indicating desire and fertility. It can be gold,

seem to decide whether it's a genuine religion using subversive humour to spread its ideals or a self-reflexive postmodern joke.

symbolizing wealth or greed; red, symbolizing desire or blood; or green, symbolizing nature's bounty. It promises eternal life, and brings tragedy and early death. It represents original sin, and salvation from sin. It symbolizes lust, temptation, beauty, knowledge, discord and danger. This round fruit that fits in the hand is a metaphorical grenade, weighed down by so many meanings that it seems impossible for one single apple to contain them all. And the more I read, the more I began to suspect that maybe it doesn't. Maybe it never did.

In many languages the orange is known as a 'golden apple'. In the Middle East, the quince is also known as a 'golden apple', as are apricots in Cyprus. In Italian, 'golden apple' translates as *pomo d'oro*, which is remarkably similar to *pomodoro*, the word for 'tomato'. When tomatoes were first introduced to northern Europe, they were colloquially referred to as 'love apples'.

The Greek botanical name *Hesperidoids* was used to refer to all citrus fruit. If the Garden of the Hesperides really was located beyond the Atlas Mountains, to the far west of the world of the ancient Greeks, say in modern-day Morocco, the fruit growing there could well have been the argan, which resembles a small golden apple. Its tree has bark that resembles reptilian scales, like those of a dragon, and it is currently feted, if not for conferring immortality, then certainly for helping maintain a youthful appearance and preventing illness and disease.

Across Europe, 'apple' has been used to describe many newly discovered fruits and vegetables, from the potato ('earth apple', still called *pomme de terre* in France) to the aubergine (brilliantly dubbed *mala insana* or 'apple of insanity' in ancient Rome). Maybe the mythological fruits in Eden and the Garden of the Hesperides weren't apples at all but lazily named crazy aubergines.

This all seemed like a linguistic mess without any clear answers, and it's led some historians to question the apple's importance. But then I read something that drew me to the opposite conclusion. Completely by accident, while I was pondering the history of mythological fruit, I came across something called the 'lexical hypothesis', an idea from psychology which states that the personality characteristics most important to people's lives will eventually become part of their language, and the most important personality characteristics will be encoded into language in a single word.

In his book *The Organized Mind* (2014), neuroscientist Daniel Levitin uses the lexical hypothesis to illustrate the evolution of language itself. Studies of thousands of languages, old and new, show that they all come up with similar principles, independently of each other. We're all hard-wired to impose structure on the world around us, and we all do it in roughly the same way.

In any language, the first distinction that is made – the root of language – is the binary distinction between human and not-human us and not-us. Further simple binary distinctions then follow: day/night, being/not being, and so on. When I first read this, I was reminded of the Book of Genesis. God created Heaven and Earth, and then said, 'Let there be light'. The next thing he did was divide light from dark, night from day, Heaven (the firmament) from the waters below (Earth.) Then he divided dry land from the water. If we apply the lexical hypothesis, Genesis reflects the evolution of language.

Levitin goes on to say that, as language evolves, finer distinctions creep in, with two or three distinctions starting to appear at once. In the classification of animals, we distinguish between things that fly, swim and crawl – birds, fish and snakes – and then mammals appear. Depending on context, there might be an idiosyncratic word for a specific species that has great social, practical or religious meaning – the whale in Genesis struck me as a perfect example.

Early languages have no single word for plants. What they do have is a single word that refers to tall, woody, growing things – trees. (Eden is planted only with trees, remember.) When language evolves to the next step, we distinguish between trees, grasses, bushes and vines. Maybe the mystery around the identity of the tree of knowledge in Eden or the golden apples of the Hesperides is simply that there was only one word for fruit trees at the time these stories were written, with the idiosyncratic exception of the fig, which, like the whale, had particular significance.*

If we then follow the rules of the lexical hypothesis and think about how language continued to evolve, further subdividing things and concepts like the branches of a tree, as it becomes more sophisticated you'll eventually start to make distinctions between different kinds of trees – say, those that bear fruit and those that don't – and eventually, trees that bear different kinds of fruit. This might explain why at one point all fruits were referred to as apples, or why at another point, a newly discovered fruit or vegetable might be referred to as an 'earth apple' or a 'golden apple'.

* The fig was associated with the liver in some ancient societies, at a time when the liver was thought to be the most important organ in the body.

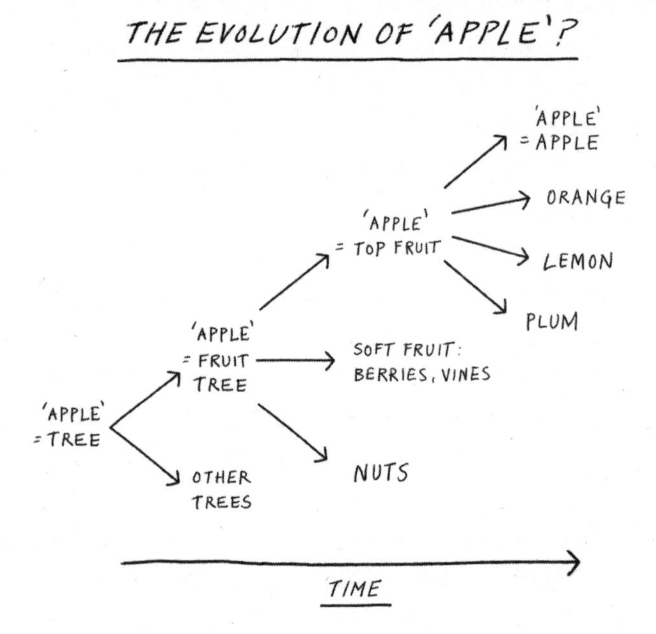

If that's the case, then the lexical hypothesis suggests that the word that eventually meant 'apple' specifically – the apples we eat and drink today – was the most important word in identifying fruit at every stage in the evolution of language. We come up with golden apples or love apples because that's the simplest way to describe new fruits meaningfully at this stage in the evolution of language. (We see a similar thing happening when the early prototype of the train was called the 'iron horse', or the automobile referred to as the 'horseless carriage'.) The confusion and uncertainty surrounding the apple's true identity in both the mythology and the reality of the ancient world fall away. The word for 'apple' is the most important word for 'fruit' in human history, and it eventually settled with the fruit that had more importance than all the rest. And that's why any non-specific sacred fruit, in areas where apples were known, couldn't be anything else. The forbidden fruit was referred to as an apple as early as the fourth century AD, and by the Middle Ages it was regularly portrayed as an apple in art.*

As well as the magic and potency and the lexical hypothesis, the symbolic apple makes mundane, practical sense. The apple is one of the most widely cultivated fruits on the planet. A serf in England or France would have easily recognized it, and would have instantly related to stories around it. It's doubtful such people would ever have seen a pomegranate or a fig in their lives, or recognized one if they did, so why try to explain what it was? Christianity, a religion born in the Middle East, had to make itself relevant in different climates and cultures if it was to survive and spread.

Exploring the apple's history leads us irresistibly to an exploration of systems of belief, of how we shape our world and make sense of it, and that in turn leads us back to the apple. It's a relationship that doesn't just make do with Christianity, but spreads wherever the apple is cultivated, is older than the Bible, and present at every aspect of the apple's cycle. Of course the forbidden fruit was an apple. Geography and climate be damned: all that matters is the symbolism. And symbolically it couldn't possibly have been anything else.

* There's also another linguistic tic that would have proven compelling to the early Christian Church. In Latin Vulgate, the word for 'evil' used in Latin translations of Genesis to describe the tree of the knowledge of good and evil is *malum*, while the world for 'apple' is nearly identical, *mālum*.

PART THREE
Ripening

July–August

'Here is the everlasting miracle before my eyes,
and all miracles are mysteries.
Once I thought I should understand such things
when I was "grown up,"
but I find myself still a boy.'

Liberty Hyde Bailey, *The Apple-Tree*, 1922

9.

Making a Tree

From a human perspective, plants are either domesticated or wild. Either they grow freely as nature intends, or we interfere with that growth in some way and guide it to suit our purposes. For most of our history as a species, we ate wild plants the way other animals do. When we learned to domesticate them, we were able to improve yield and consistency, which freed us from a subsistence-focused existence and allowed us to create towns, universities and libraries, and put cute pictures of cats online.

The early history of agriculture contains three great leaps forward in the domestication of plants, each more sophisticated than the last. First came the selection and cultivation of seeds. Grains and pulses can self-pollinate. If they do, they'll grow true to type. In other words, you'll get very similar crops from successive generations of seed. Around 10,000 to 12,000 years ago, early agriculturalists began collecting and sowing the seeds of wild grasses such as emmer and einkorn wheat, and pulses such as lentils and chickpeas, selecting those that grew well and propagating them. They weren't interfering with the structure or identity of these plants, but they were growing them in a domesticated way, constantly improving the quality of favourable varieties by a principle, essentially, of unnatural selection. This innovation made food supplies more secure, and played a key part in creating the basis for fixed, permanent settlements as opposed to nomadic drifting. Something as fundamental as the birth of civilization doesn't have one cause but a cluster of them; among that cluster, the birth of agriculture, from which bread and beer flowed (though not necessarily in that order) is a pretty fundamental one. The first cities followed soon after.

Unlike grassy plants, most useful woody plants, such as the apple, are heterozygous – they usually need to cross-pollinate to reproduce, and therefore don't breed true to type from seed. In the fourth or fifth millennium BCE – thousands of years after we had the first farms growing the first crops outside the first cities – we get the earliest evidence of agriculturalists cloning favoured species of fruit by artificial, asexual propagation. Some plants, such as figs, pomegranates, grapes and olives, will sprout roots quite easily from cuttings. This can be done by simply planting cuttings in the ground and waiting for roots to sprout, or, more reliably, by the practice of 'layering', where soil is built up around a stem or branch until roots grow from it, and then cutting it from the parent and planting it. In this way, the first vineyards and olive groves were propagated.

But other woody plants, such as pears, plums and apples, don't root so easily.* There seemed to be no way to propagate

* That's not to say they don't root at all. Sometimes the canopy of a tree will brush against the ground and new roots form. What were once branches become a secondary wave of trunks that establish canopies of their own. There was a famous perry pear tree on the estate of Lord Scudamore at Holme Lacey in Herefordshire that once covered most of a hillside, and parts of it still survive.

from a favoured tree until somewhere between 1000 BCE and 500 BCE when, possibly inspired by the natural grafts trees sometimes make when they rub together over a long period of time, agriculturalists worked out methods of fusing together the parts of two different plants in such a way that the characteristics of the top one could be spread to new roots.

No one knows where and when this practice of grafting was initially perfected, but it's almost certainly responsible for the apple making its way from Asia into Europe. The first written evidence of it starts to appear in Greece around 500 BCE, possibly after having been brought back by Alexander the Great – along with the apple itself – from his adventures in Persia. Grafting is discussed in 'On the Nature of the Child', thought to have been written around 420 BCE by one or more followers of Hippocrates, and goes into great detail, suggesting perhaps that the practice was already well known by this date:

> Some trees, however, grow from grafts implanted into other trees: they live independently on these, and the fruit which they bear is different from that of the tree on which they are grafted. This is how: first of all the graft produces buds, for initially it still contains nutriment from its parent tree, and only subsequently from the tree in which it was engrafted. Then, when it buds, it puts forth slender roots in the tree, and feeds initially on the moisture actually

This phenomenon in all likelihood is the inspiration for supernatural forests that 'walk', like Birnam Wood in *Macbeth*.

> in the tree on which it is engrafted. Then in course of time it extends its roots directly into the earth, through the tree on which it was engrafted; thereafter it uses the moisture which it draws up from the ground . . .

This understanding is, of course, imperfect – the graft doesn't put its roots down through the tree. But early thinkers believed that the moisture mentioned in the passage above gave each plant its individual characteristics. They called it 'specific fluid', and believed each type of plant drew up a particular fluid from the earth. This belief was still popular in England as recently as the seventeenth century.

These days we know much more about how grafting works and why. But that knowledge has done little to change the technique used 2,000 years ago. And on a shining August morning, when I almost bunk off and go to the beach instead, I get to learn how to do it.

At this time of the year the orchards look stunning once more, as the fruit starts to swell on the branch where the blossom sat two months before. The orchardist knows whether there will be fruit to harvest or not – the tree has finished shedding any fruitlets that haven't really taken, and the risk of fruit being killed off by late frost has passed. Now it's a case of protecting the growing apples from all the pests that find them so attractive and the diseases to which they are so prone, and hoping that the combination of temperature, sunlight and rainfall keeps them growing bigger and juicier all the way through to harvest time.

It's a Saturday and I should still be in bed. Instead, I'm changing trains at Brighton for the ten-minute connection to Falmer, on the city's outskirts. The sky is that rare blue that's solid and endless, with nothing, no clouds or contrails, disrupting it. A fifteen-minute walk from the station, across the perpetually busy A27 dual carriageway, and past the University of Sussex, and I come to Stanmer Park, former seat of the Earl of Chichester. The estate comprises over 5,000 acres and contains the villages of Stanmer and Falmer; it's also a nature reserve that's home to a large bat population, the Low Carbon Trust and the Earthship Brighton, an innovative new type of building designed to operate entirely off grid, generating its own electricity and processing its own waste.

A few people are already waiting when I arrive. We're all ages, men and women, and various nationalities. There are a fair few dreadlocks. Eventually Stephan, our chaperone for the day, collects us and ticks off our names. Stephan Gehrels works for the Brighton Permaculture Trust, a charity that promotes greener lifestyles and sustainable development, following the principles of permaculture: that the best and most sustainable ideas for living are those that mirror natural ecosystems as closely as possible. Rather than trying to improve nature, the idea of permaculture is to cleave to it and work with it, following its contours and rhythms.

The Brighton Permaculture Trust puts these principles into action at various sites around Brighton and Sussex, with most of its attention focused on a plot in Stanmer

Park. Stanmer Organics is a collection of buildings, gardens and allotment-style plots that are leased by various environmental organizations. It includes three orchards, two of which were long abandoned and are being nurtured back to health by the Permaculture Trust. This is where they hold courses on how to look after fruit trees, as well as teaching and practising permaculture in both agriculture and building design.

The gardens are outlandishly beautiful. The size of the plants – very few of which I recognize and some of which seem to belong in a jungle – suggests that permaculture principles are successful. There are rows of greenhouses, thistles that are taller than me and, possibly a triffid.

We're left to make tea and small talk for twenty minutes, because this has to be built into the schedule of any organized group event of any kind in England. One man – tall, blond, bearded and shaggy-haired – wanders over to me.

'Come far?' he asks.

'Stoke Newington, North London,' I say, 'on the early train.'

He nods. 'I'm Wiccan,' he says.

As an opening conversational gambit, this announcement takes me aback. I expected people here to have alternative lifestyles and beliefs, but it seems a little sudden to share them in this way. Is he going to try to convert me? Do Wiccans feel they need to warn people when they meet them in case they suddenly start praying to the trees or something? Does he think that because I live in Stoke Newington I might be Wiccan, too? I think about saying something like 'Good for you!' or 'Oh, I've been looking at the revival of alternative belief systems, what's that like?'

But anything I can think of sounds too forward, even though he over-shared first. I have absolutely no conversational comeback, so after a few seconds of awkward silence I wander off on the pretext of making another cup of tea.

Just as I'm looking to make sure I put my soggy tea bag in the right composting bin, Stephan announces that we're heading over to the classroom. I find myself walking behind the tall, blond Wiccan, now introducing himself to a grey-haired, rather prim lady. She's telling him she lives in Brighton.

'High Wycombe,' he replies.

We take our seats in a small classroom lined with shelves of books and pamphlets on horticulture. At this point it's like a treasure trove to me, and I struggle to keep my attention on the front of the room. Peter May is a horticulturalist who has propagated over a thousand fruit trees in Sussex. He looks very different from most other men in the room; he's a gentle, softly spoken man with a careful, professor-like bearing, the kind of peaceful, low-volume person who makes you lean forward to hear him better rather than miss anything. He asks us to introduce ourselves and explain why we're here. Most people have either acquired or want to acquire a small bit of land and move some way towards being self-sufficient. There's a definite desire to 'help nature fight back', but at the same time the self-sufficiency part feels like it's helping some of these people fight back too, or at least become less reliant on a society and a system they don't respect and don't want to be quite so enmeshed in.

Peter is going to teach us about a method known as chip-bud grafting. That's because, compared to other methods, it's quite easy for beginners to pick up – if not to master – and has a high success rate, especially with apples, where it's almost 100 per cent successful if you know what you're doing. It's the main method used in commercial nursery production to create new fruit trees and is popular for apple, pear and plum, as well as ornamental species such as cherry, ash, maple, lime and hawthorn.

At its simplest, there are two parts to the graft: the grafting wood, from which you take a bud, and the rootstock on to which you graft it. With the method we're using today, you cut out a chip containing a bud from the grafting wood, cut a chip that's very slightly bigger into the side of the rootstock, and tie the two pieces together until they have time to bond. If the graft is successful, your bud will grow, and it will fruit according to the characteristics of the grafting-wood tree rather than the rootstock. Sounds easy, doesn't it?

The top or graft wood is usually referred to as scion wood. I love this word used in this context. When you're talking about people, a scion is a direct descendant of a notable family, the heir to the fortune and title. In modern usage it's sometimes chosen to suggest other characteristics – 'heir' seems more passive, someone who is waiting to inherit, whereas a 'scion' is the hot-blooded tearaway who needs to settle down before the title passes to him, or else he'll squander the family fortune on women and gambling. Prince Charles is an heir. His Mills & Boon equivalent is a scion.

You can buy scion wood from the National Fruit Collection at Brogdale Farm in Kent, or you can collect it from orchards or in the wild yourself if you know what you're looking for. We don't know, but Peter and Stephan do, so we go for a walk in one of the old orchards.

We follow a path past a greenhouse so old the windows themselves have taken on a chlorophyll hue, and then head into a walled enclosure that was once the orchard for Stanmer House. It reminds me of fairy-tale castles that have been neglected for a hundred years, where nature is reclaiming the walls and paths. The Brighton Permaculture Trust is gently pushing back after decades of neglect, not clearing out and cutting everything away, because that's not what they do, but working with the trees, making them healthy and encouraging them to be productive once more.

Chip-bud grafting has to be done between the last week of July and the start of September. I always had a very non-specific understanding of the passing of the seasons before I became interested in apple cultivation. When we have Indian summers and mild winters, or when the groundhog predicts there'll be six more weeks of winter or an earlier spring, we can be forgiven for thinking that, while the seasons come in a fixed order, the timescale varies. I'm surprised to learn that many elements of the seasons are fixed, even if the weather they bring is increasingly variable. The aspect that never alters is day length, and that has a huge influence on plant growth.

You need to use first-year growth for your scion wood, and it needs to be starting to develop a woody base. It can't be too soft and whippy or it will rot, so you can't start before the end of July. And when you've successfully completed your graft, you need six to eight weeks to make sure it has taken. This has to happen in the growing season, and growth stops at the end of October, after which the tree falls dormant. So you can't graft successfully after early September, and that makes August a time of hard grafting.

Once Stephan and Peter have shown us how to spot first-year growth, it's easy to pick out, and I marvel at how vigorous a healthy tree is. Here, in early August, the old apple trees are showing a good eighteen inches of growth since the end of winter. On most branches there's a small knot, a slight bulge, and below this the bark is of a darker hue. Little suckers sprout all along the trunk, which will grow into new branches and boughs if they're allowed. Most of the growth is up in the canopy, where the light is best, but trees don't take any chances. When they're being cultivated, you usually prune the suckers to keep all the energy going into the branches above that have already been shaped and are already productive.

Having selected your scion wood, it's important not to let it dry out, so you cut off all the leaves, leaving a tiny bit of leaf stem to minimize the damage to the bud to which the leaf was attached. You then keep your wood in a damp bag in a cool place until you need it, refrigerating it if you're not grafting it the same day.

With our scions wrapped in plastic bags, we return to the classroom to learn about rootstocks. The rootstock determines how fast and how big the grafted tree will grow, and it needs to be from a plant closely related to the scion, though not necessarily the same species.

With a rootstock in the ground and some scion wood nearby, the next thing you need is a grafting knife. A few weeks before the course, I bumped into Stephen Wood, an apple grower and cider-maker from Vermont who is largely responsible for reintroducing cider apples, and the expertise of cultivating them, back to the USA after almost a century's absence. We were in Bristol, at a cider-industry event, and I told him proudly that I was going on a grafting course. 'Ah, don't go on a *course,* come and see me instead!' he said, pulling a small knife from his pocket. It had a plain, weathered wooden handle and a small, grey blade, and looked as if it had been used so much it was moulded to his hand. It was his grafting knife. He was in England on holiday, but he still carried it with him. It struck me then that a grafting knife must be a very personal tool for an orchardist, like the individual roll of knives belonging to a chef.

Our course instructions recommended that we bought our own grafting knives as they couldn't guarantee they would have enough for all of us. After carefully examining the options on Amazon, I selected a quite expensive one with a beautiful wooden handle and a curved blade that resembled a miniature scimitar or baling hook. This would be my grafting knife, and one day it would be as moulded to the shape of my hand as Stephen Wood's was to his.

It's the wrong type of grafting knife. As Peter May takes us through knife choices and knife care, it turns out mine is designed for March grafting, which is a completely different technique from chip budding. For this we need short, straight blades that we can control carefully to make neat little cuts, and mine is too big. But Peter says it's still a good knife, and I should probably be able to manage.

Peter demonstrates the process five or six times before we go outside on to the lawn to practise for ourselves. He makes it look easy, of course. First, you take the scion wood and cut off all the leaves carefully with secateurs, leaving the buds. You choose a decent bud and slice into the stem just over halfway through, about 8mm below the bud.

Supporting the bud with your finger, you then start making a second cut an inch and a half above the bud, slicing down smoothly and carefully towards the first cut to free a sliver of bud framed by bark and the subcutaneous layer of cambium behind it.

Turning now to your rootstock, you need to make a new home for your bud. You make a sharp nick in the stem, slicing down at a 45-degree angle. An inch and a half above – slightly taller than the size of your bud chip – you make a second cut, slicing down towards the first and chipping away to reveal an arch-shape with a slight lip at the bottom. It's important that the sides are parallel and smooth – if they're rough, woody fibres will remain in the graft and rot, and there'll be a gap between the two parts. You match up the bud with the opening on the rootstock, cambium to cambium, ideally with the bud resting on the little lip. In the cambium layer, cell division occurs quickly and the two parts should grow and bond together. Until this happens, you wrap the whole thing tightly in tape to keep the two parts close together, leaving the bud itself free and exposed to the air, but bonded all around. Six to eight weeks later you see if the graft has taken or not. If it has, you cut the tape away and, the following spring the bud should start growing.

Out on the lawn, which is fringed by giant thistles, and with the sun beating down, Stephan has sunk a dozen lengths of willow into the grass to simulate rootstocks. We pick out our scions and get to work.

My first attempt is pretty good. I'm in control of my knife – slow and gradual – and I create a nice bud chip. My first go at cutting a sliver from the rootstock isn't quite right – it's too tall, meaning the cambium from the bud will sit inside that of the rootstock. My second attempt is pretty good: a tall, white arch with parallel sides framed by the soft lime green of the cambium: a perfect halo. My chip sits on the lip at the bottom and stays in place when I take my fingers away. I tie the tape tightly, leaving the bud exposed, and I've made a tree! Well, I would have if this were real rootstock and not a dead stick driven into a lawn. I think about what I'm doing, trying to centre myself in the moment. And that's when it all goes wrong.

While trying to make my second chip bud, the wood offers up more resistance. I have to gently rock the knife to force it through, and push the wood into my stomach to steady it. The knife is very close to my gut with my thumb over it and my fingers are in the way. My self-preservation instincts tell me this is all wrong. I hack at the wood, scared of it, as if I have to defeat it. I produce rough cuts and split stems. Suddenly, I hate this. And it doesn't help when other people begin to cut themselves. Soon, Stephan is fully occupied by handing out sticking plasters.

Peter sees me struggling and comes across to help. He positions my hands correctly. 'You have to get close up,' he says. 'You have to think about what you're doing with your knife all the time.' But looking down at my hands, I feel a familiar clumsiness creeping up on me, a sense of impending ineptitude. I've always been rubbish with my hands. At school I was bad at sport, and have the mental scars to prove it. At home, I'm pretty useless with DIY. And now I'm standing in a garden with a very sharp knife angled towards my stomach.

I have to stop and try something new. I feel daft about

this, but maybe being among people who are certainly spiritual and definitely of an alternative bent helps, and I improvise a mental exercise. Instead of looking down at my stupid hands from above, I try to move down into the hands themselves. I try not to be in my head any more, but actually in my fingers. I suppose all I'm really doing is trying to shift my emphasis from sight to touch and feel. But it helps. I become steadier, more patient. My breathing slows and I relax, feeling the knife become part of me rather than my enemy. Gently, slowly, I begin to make smooth cuts, and it doesn't feel dangerous any more. Each time I make a cut, the initial crunch of the first bite into the bark feels good. I think I hear it, which I don't. I'm feeling the vibration of it, and I focus on that feeling.

Next, we have to cover our tentative union with grafting tape. I find this part more difficult again. The most important thing is not to cover the bud itself to keep it open to the air as you seal the sliver of wood around it, taping it tightly to keep the two layers of cambium pressed together. You have to keep the tape absolutely tight at all times, crossing it below the bud and tying it off. At first, no matter how hard I try, my tape always slides over the bud, but I persevere. After half an hour my grafts are not uniformly perfect, but I know – and feel – what I'm doing.

We break to eat our packed lunches in the sun, and I use the time to find out more about what the Brighton Permaculture Trust is doing with apples. Stephan explains that they're planting trees wherever they can in Sussex, working with kids, and especially in less affluent areas. 'If you can encourage a tree to grow on a council estate and the people to look after it, then you're changing their ideas about food. You get a good harvest and they're participating in local food production and getting nutritious food for free,' he says.

Sussex is not often cited as a centre of apple-growing, but there's a great deal of heritage here, with many varieties clearly related to the county. 'The localities, the people and the apples all interrelate,' says Peter, and I'm reminded of Jackie Denman in Putley saying almost the same thing. *It's made by somebody. It comes from somewhere.*

Peter and the Trust are trying to track down the people who first propagated successful varieties locally and make a record of them. 'Take the Bossom apple,' says Peter. He explains that this was first identified as a local variety by Robert Hogg in 1851, but by the 1980s it was down to just one known tree, growing in a garden in the village of Graffam. In the gale of 1986, that tree fell over, but people took grafts from it, sent some to Brogdale and to the Frank Matthews tree nursery, and now it's growing in Stanmer, too, no longer endangered.

Does preserving such varieties have any value beyond a sentimental attachment to local identity?

'Maintaining genetic diversity is so important to the future of the apple,' says Peter. 'For example, you have to look at building resistance to scab and canker, and creating obvious safeguards for the future. Concentration on fewer varieties leaves us more vulnerable; we lose traits and characteristics that would help apples survive in different conditions. There's a natural element to it, just like when we look at the human gene pool. It's not just a whimsical romance for the trees of the past.'

I think of the huge gene pool discovered in the apple

forests of Kazakhstan, and how the optimism and excitement of it is tempered by the fact that 80 per cent of it has already disappeared. We may have only 20 per cent of the genetic diversity present at the apple's birthplace, but surely that's an argument for holding onto as much of it as we can.

We've learned the method, so after lunch we practise, which Peter urges us to do again and again when we get home. But there's only so much you can do in one session, so eventually we sit on the grass while Peter teaches us other approaches to the same idea. There's T-bud grafting, where instead of cutting a chip out of the rootstock, you slice a T-shape into the bark to expose the cambium, pull the flaps aside and snuggle a slightly bigger chip bud inside the incision. With whip-and-tongue grafting, you cut all the way through the scion and the rootstock with a lightning shape that hopefully allows the two to fit together like jigsaw pieces, and then there's March grafting, where you use a knife like mine to make one sloping cut all the way across the rootstock, and a corresponding slope across the bottom of the scion wood, and tie them together.

When you've finished a graft and you know it's been a success, it's standard practice then to prune away any surviving rootstock above the graft so that the tree grows true to the scion wood rather than as some kind of hybrid. But you don't have to. If you left the original stem growing as well as the graft, you'd have a tree with two different varieties of fruit on it. I catch Stephan grumbling in the background about some 'abomination' of a tree with hundreds of varieties on it. To be fair, such a creation sounds about as far away from the principles of the Permaculture Trust as you can get, but I'm still intrigued. A quick internet search reveals that Paul Barnett, a horticulturalist living in Chidham, near Chichester, West Sussex, has spent the last twenty-four years cultivating a tree in his back garden, carefully adding new bud grafts every summer and winter grafts every year. This single tree now bears over 250 different varieties of apples. When asked by a newspaper journalist why he did it, Barnett replied, 'It's really important for people to know what kind of apples they are growing . . . There have been some varieties which have been lost over time. I don't want to see any disappear. You don't know what will happen in the future with global warming or pesticides. You may need to crossbreed apples with older varieties to make them resistant to such things. That's why every type of apple is worth preserving.' So his aims are exactly the same as those of the Brighton Permaculture Trust, even if his methods are radically different.

Another group of quirky grafters sound as if they'd be much more likely to gain Stephan's approval. Early in 2015, 'guerrilla grafting' hit the streets of cities like San Francisco. These streets are full of pretty, ornamental fruit trees that look nice but don't bear any fruit. Fruit, as far as city officials are concerned, falls on the floor and makes a mess. It rots and stains and has to be cleared up, at considerable expense. But the Guerrilla Grafters, a 'self-selected international cultural workforce' whose motto is 'Undoing civilization one branch at a time', want to provide free fruit for the homeless of the cities in which they operate. They work at night, illegally grafting scions from fruit-bearing trees on to ornamental rootstock. The genius of their approach is that the grafts will take three to four years to bear fruit, so they'll be well established by the time anyone knows how many grafts they've made, or where they've made them.

Grafting gives you such great power; there's something quite intoxicating about it. You can make a tree! You don't just have to grow it from seed and hope for the best, you can design it to your own specification. The question is, do you use that power for good or evil? Do you preserve old varieties, wipe them out by propagating a few popular ones, or create freaks of nature? Of course, it's more nuanced than good or evil – everyone employing grafting techniques believes they're doing the right thing. And it's hard to imagine anyone opposing the practice in principle. It might sound as if you're trying to play God, but it's one of the founding principles of agriculture – and, in turn, of civilization itself.

10.

Fruit Focus

East Malling in Kent is one of those pretty countryside villages where the ancient church steeple is still the highest point around. The noisiest thing that ever happens here is the train rumbling past on a high bank at the top of a modest hill. I alight from the 9.52 a.m. service from Victoria on to an empty platform, from which it's a gentle walk downhill and through the deserted streets.

One of the best parts of my job is that I'm lucky enough to travel against the flow. It's obvious that East Malling is a dormitory village: apart from the houses, there's one pub, a hairdresser's and an Italian restaurant, and I don't meet a single other person as I walk through the village. The people who sleep in these houses are on the overcrowded, overpriced trains into Victoria every morning, and are back fighting to get on those trains every evening. Living in London and writing about aspects of the countryside, my trains are always empty. Every time I witness the crush and look at the faces of the people going the opposite way to me, I count my blessings and hope that living somewhere as peaceful and green as this makes it worthwhile for them.

Opposite the pub at the bottom of the hill, I turn down a long private road stretching away through fields and orchards. Here, at the height of summer, the fruit is fully formed and the crop looks good. The orchards to my left are full of shining red apples on dwarfing rootstock, and resemble row upon row of decorated Christmas trees. To my right is a seemingly random jumble of different varieties. Fruit the size of golf balls, fruit the size of tennis balls, every shade and stripe.

These trees belong to the East Malling Research Station, the biggest and most influential horticultural research organization in the UK. Its roots go back to 1912, when M. J. R. Dunstan, the principal of Wye College (the School of Agriculture within the University of London, based in the village of Wye in Kent), addressed a meeting of 600 fruit growers in Maidstone on the subject of 'The scope of scientific research in fruit growing' and argued for the need to establish a fruit research station in Kent. The Board of Agriculture, Kent County Council and the growers themselves funded the purchase of 23 acres of land at East Malling, and the Wye College Fruit Experimental Station was founded in 1913. It became the independent East Malling Research Station (EMR) in 1920. For a hundred years, EMR has researched all aspects of horticulture, focusing mainly on fruit, and now offers consultancy services to growers and industry bodies, including DNA fingerprinting to help accurately identify and describe varieties, crop protection trials and trials in 'growing media' – the composition of soil.

Today, East Malling is hosting Fruit Focus, which describes itself as 'the Industry's Premier Fruit Event'. Each year, around 120 exhibitors set up shop in a couple of East Malling's fallow fields to showcase the latest developments in every aspect of fruit growing, to a crowd of around 1,300 fruit growers and industry professionals.

Jon Day, event director at Haymarket Exhibitions, writes in the introduction to the day's programme, 'Fruit production techniques are moving at a rapid pace. This year's attendance illustrates the value that top soft and stone fruit growers put on the event, which offers an unrivalled opportunity to stay abreast of rapid change, keep businesses efficient and competitive, and to network with other progressive growers and leading industry figures.' Nigel Trood, managing director of principal sponsor Mack, adds, 'We always enjoy being at the heart of Fruit Focus. This year we placed particular emphasis on quality and advances within the technical sector' – as if there's ever a year when an industry goes, 'You know what? We're not really bothered about quality and technological advances at the moment.'

As with any pursuit, if you don't grow fruit for a living, you have absolutely no idea how many aspects you need to look at and how many things you can buy. A typical exhibitor is demonstrating 'ongoing developments in its soft-fruit-tunnel monitoring systems to improve irrigation practice and disease management using telemetry and software', as well as a new fire-blight early-warning service 'enabling preventative or antagonistic sprays to be applied more accurately' and 'novel mapping and scanning techniques produced by drones to better manage crop variability'. There are seminars on aligning research funding to help secure growth and raising the bar on quality, press briefings on controlled-atmosphere preservation and the optimum substrate for growing blueberries, and companies with inevitable slogans along the lines of 'Sowing Ideas, Reaping Success' and 'Plant the seeds and watch your business grow'.

I find it all a little bewildering, and quite alienating. I don't know anyone here, and it all feels a long way from grafting at Stanmer or walking through the blossom at Dragon Orchard. I want to gain a new appreciation of nature and reacquaint myself with the slower, deeper rhythms of the world, to pay attention to the wheel of the year and the cycle of the seasons. But now the apple seems to have pulled me back into the world of corporate banality I used to inhabit and have tried to escape. I realize that fruit growers such as Norman Stanier probably understand all this much better than I do, but I doubt they want to be talked at like this. I fight the urge to leave, persuading myself with the aid of a cup of tea, a bacon roll and a seat in the weak, hesitant sunlight to sit and go through the programme and take in my surroundings.

As I read through the full list of exhibitors, I start to build a picture of the most pressing needs of modern fruit growers. Working on the assumption that the more companies there are trying to sell something, the more demand for it there must be, the priorities for the modern industry are the recruitment of seasonal labour, the latest irrigation techniques and, most of all, the constant battle against bugs and pests.

Since I started writing about cider, I've been to quite a few country shows. Bill Bradshaw and I are often invited to do book events or judge cider competitions, and the sheer scale of the shows is often overwhelming. A country show is two entirely different events coexisting in the same physical space. It's a trade show for farmers, similar to Fruit Focus, offering all the equipment and advice they might need. But it's also a festival for the general public, celebrating all aspects of country life. Throw into that the random and opportunistic retailers who always set up wherever there are large crowds, the ferret racing and stunt motorcycle displays, and the result can be a jumbled bazaar that verges on the hallucinogenic.

At the Royal Bath & West, the biggest country show in the UK, which takes place in May, I strolled down one single avenue of stalls that included stalls selling ultra-sharp kitchen knives, leather goods, a 'cheese village', sit-on lawn mowers, a double-decker bus with a sign outside that read 'COME IN FOR COFFEE AND????' which, on closer examination seemed to be selling insurance; a sweet shop, and a display warning me in stark terms about the dangers of the countryside.*

Visit the Bath & West show after winning the lottery and you'd be tempted to go on an impulsive spree just to see the reaction, to see if you could actually get away with saying, 'OK, I'll take one of those sit-on mowers, some Caerphilly, three sets of pans, a pre-formed concrete drinking trough for cows, a Jacuzzi, some jelly babies, a novelty dog harness, and a combine harvester please. Could I have some help with packing?'

Next door to 'Cow Art', which sold a more diverse range

* Neither Wicker Men nor duelling banjos were featured.

of objects with cows on than you could ever need, was Upsi-Daisy, a company selling specialist cow-lifters. When cows fall on their sides (or are tipped over by stoned American teenagers in Hollywood movies), they're unable to get back on their feet again without being assisted. At least, that's what I thought. The Wikipedia page on 'cow tipping' claims this is an urban legend, and that cows are perfectly capable of getting back on their feet. And yet there we were: unless Upsi-Daisy is a very elaborate hoax designed to perpetuate an urban myth, a safe lifting harness is an important tool for any livestock farmer.

The cow-lifter could be used to transport livestock, lifting erect cows on and off lorries. But the Upsi-Daisy stand had a video (not a live demonstration, thankfully) demonstrating the apparatus being used to help lift a prone cow back onto its feet.* A casually dressed man and his three small children watched in fascination, forming a chain with their hands, their eyes wide. The proprietor of the stall, grey mutton-chops sprouting magnificently from beneath his flat cap, approached them and asked, 'Dairy or beef?'

Clearly flustered at being approached, the father replied, 'I'm sorry?'

'Are you *dairy* or *beef*?'

'Oh! No, I'm neither. I'm not a farmer. We're just here for the day and we were just watching the, er . . .' He pointed at the video.

'Oh, 'e just wants a look at 'er, does 'e? Fair 'nough, ain't

* If you think I'm making this up, you can see this video at http://upsidaisy-cowlifter.com/

nothing wrong with thaaat,' said the older man, and went back to his stool.

I've always found that looking at huge pieces of farming machinery is a good way of settling down and feeling more grounded at events like this. The Bath & West boasted behemoths that must surely be called super tractors; extreme, bulked-out-and-fucked-up-on-steroids farm machinery. The only use I could imagine for them was that any man suffering from feelings of inadequacy could get behind the wheel of one, fire her up, drive ten yards, then drop from the cab and say gruffly, 'Yeah, that's done it. I think I'll be OK now.'

Unlike the Bath & West, there's no public element at Fruit Focus, but the variety of stalls is no less daunting. So, before I do anything else, I head for the big machines to get my head straight. The new range of tractors on show here (from John Deere's latest 5G series) are much smaller and more delicate than the titans of the Bath & West, but the array of spindly bits, tanks, hoses and hydraulic limbs is more diverse than any other I've yet seen. These look like very expensive bits of equipment with unfeasibly specific uses. It reminds me of *Thunderbirds*, when, in each episode, the big cargo ship *Thunderbird Two* selects which pod it needs for each mission. Pod Four carried *Thunderbird Four*, a miniature submarine, and Pod Five carried the Mole, a subterranean vehicle with a massive drill on the front. (Whatever was in the first three pods wasn't as memorable.) I was enraptured by this show, and the only point at which my suspension of disbelief wavered was when the only thing that could, say, rescue a passenger jet that would explode if it lowered its undercarriage was a set of heavy-duty wheeled drones that could match the speed of the jet and take its weight when it came into land and – oh, look! – the Tracy brothers just happened to have the very things lying around in one of the pods. I simply couldn't believe that even the coolest rescue organization in the world would have such a wide range of equipment, each bit with such a specialized use. But they had nothing on the range of gadgets available to the modern fruit farmer. I wonder if some tiny part of this is driven by a hitherto unacknowledged *Thunderbirds* fandom among fruit farmers that rivals my own.

In the demonstration arena, one of these pieces of equipment – a baby crop sprayer – is being put through its paces on a few rows of apple trees that have been planted specifically for such demonstrations. I have no idea what I'm looking for – uniformity of coverage, perhaps? – but it seems very impressive.

Cheered up by the big toys, I make my way to the English Apples and Pears stand, where there's a display promoting the Bramley apple. I learn that Bramleys are used in hundreds of commercial products and receive the endorsement of celebrity chefs. They have a 'tangy taste and fluffy texture that's perfect in both sweet and savoury dishes'. The Bramley campaign was set up in 1989, and was followed a year later by the formation of English Apples and Pears, a trade association that, as the name suggests, promotes English-grown apples and pears and safeguards the interests of English growers. Its chief executive is Adrian

Barlow, a smart, suited middle-aged man who combines the demeanour of a CEO with a breathless passion for his subject more typical of a beer geek or film buff. I tell him about my journey through apples and orchards so far, about the romance of it, the work being done to preserve traditional varieties, and ask what the priorities are from his perspective, what I'd call the corporate, industrial point of view. It instantly becomes clear that he feels I'm missing a big part of the story.

'Yes, the old varieties are nice, but they simply don't meet the standard required today. And if we don't meet that standard, we don't get listed in the supermarkets that control 85 per cent of food sales – they buy shinier foreign varieties instead,' he says quickly. 'We've done an awful lot of hard work to wrest market share back from these foreign varieties and persuade people to choose English apples instead. We've convinced people that English apples taste better – *which they do*, better than the same varieties grown elsewhere – and persuaded them to look for the Union Jack symbol.'

Barlow has presided over a wholesale revival in English apple sales. Back at Stanmer, Peter May had told me that joining the European Union in 1973 had tipped the industry into decline. There had been an informal agreement that apples from other countries wouldn't be imported until after Christmas, and when that agreement was broken, foreign varieties undercut English prices. Varieties such as Golden Delicious and Granny Smith received aggressive marketing support and stole the market from traditional English varieties such as Cox.

In many product categories across many countries

around the world, we tend to assume that imports are inherently flashier and more interesting than the dull domestic goods with which we're familiar. The apple is no exception. But Barlow educated the trade, the supermarkets and the apple eater about the importance of climate, how (normally) steady rainfall and stable temperatures allow our apples to grow relatively slowly and therefore develop their full flavour potential. He's promoted the characteristics of specific varieties that may not be unique to England, but, he maintains, grow better here than anywhere else. The star of the show has been Gala, which went from selling 10,000 tonnes in the year 2000 to 50,000 tonnes in 2015. In 2002 the Jazz apple was planted in trials, and went into full commercial production in 2007. A cross between Gala and Braeburn, Jazz is hotly tipped as the next rising star.

But it does seem strange for an English-apple advocate to be celebrating these varieties: Jazz, Braeburn, Gala and another market share champion, Pink Lady, are all cultivars that were originally raised in New Zealand. Barlow is a pragmatist. 'These are the varieties people want. They're crisp and sweet, with a good crunch and a nice skin texture.'

I have to admit, the bi-coloured Jazz, crimson red over a pale yellow background, and the dappled peachy blush of Gala, do look wonderful. Barlow argues that traditional English varieties such as Egremont Russet and Cox's Orange Pippin simply don't look as attractive in our increasingly visually led society. And just as importantly, they don't have as high a yield. If English growers want to make a living from apples, they have to grow the varieties that people want, hence the message that while these varieties may have originated from down under, they grow here better than anywhere else. One in four apples sold in Britain is now Gala, with a revived Cox holding a similar share. Add in Braeburn, and three apple varieties make up over 70 per cent of our total eating-apple market. That's alarming to anyone who cares about the diversity of the apple, but the consolation for the British industry is that all three are mainly grown here, with imports of Granny Smith and Golden Delicious having now been pushed well down the rankings.

I like Adrian Barlow. The promotion of English apple varieties is clearly a labour of love for him, whatever else it may be. And although he's responsible for promoting an industry, he doesn't use the language of trade shows and press releases – he still says 'people' rather than 'consumers' and, among those who work in any kind of marketing or promotion, that makes him a rare breed.

Yet I can't help being depressed by the conversation. It seems to be a recurring theme: economic growth and the potential offered by globalized markets are supposed to make us more prosperous. Perhaps in our role as consumers we are: we have way more choice than ever before, and the prices of many goods are stable.

But the flipside is that everyone has to work so much harder to keep a stake in the game. Every freelancer or self-employed person I know, in any discipline, is working harder for less money than they were ten years ago. My original career, advertising, used to have a reputation for outrageous excess which had passed by the time I started, but the occasional long lunch and a decent Christmas bonus were still regular perks for working sixty-hour weeks on a contract that assumed you worked forty. Now the working weeks are eighty hours long, young ad execs don't believe me when I tell them we used to get bonuses, and lunch, breakfast and often dinner are eaten out of a paper bag or plastic carton at the desk. And all these people are lucky compared to those on zero hours contracts and unpaid internships, who have to suck it up and console themselves that at least they're not driving growth and creating prosperity by working in sweatshops under slave labour conditions.

Sales of English apples are growing, but the overall apple market is declining as people switch to junk food. Among the good-news headlines, recipe ideas and nutritional factsheets on the English Apples and Pears website are warnings that 'Your Apple Industry Needs You'. There's a lack of appreciation for growers whose margins are constantly under pressure and who need to endlessly improve their yield per acre. Economic growth suggests we should be feeling more prosperous. But the price for that is we must compete ever harder. We must be more businesslike in everything we do. And that means there's no time to be sentimental about things like old apple varieties.

So I leave the English Apples and Pears stand with conflicting feelings of excitement and disappointment, the same discord I felt when I saw my football team, Barnsley FC, get to Wembley for the first time in their history, play woefully and lose. You want to be happy and you want to be devastated, and you're both and therefore neither. I see the corporate stands in a different light now. I still don't like their tone, but they're an army of support services that the fruit farmer must engage with if they want to compete, always looking for that extra edge. Eden feels further away than ever.

11.

Down in the Dirt

A couple of months later, I'm back at East Malling with a string of appointments.

My visit to Fruit Focus ended on a happier note when I found East Malling's own stand there. That's where I learned that 2015 was the International Year of Soils, a fact that cheered me up enormously. To celebrate, the EMR marquee was filled with exhibits that could succeed in making anyone fascinated by and excited about dirt.

What I love about the scientific method employed over the years at East Malling is the apparent simplicity of it. One display consisted of a block of soil that was excavated from dry heathland south of Tunbridge Wells by Mr Basil Furneaux, a scientist at EMR, in the early 1930s. This three-foot-deep cross section was deemed so fascinating that it was included in exhibitions and shown to the Duke of York when he visited the station in 1934. Over 2,000 years earlier, Xenophon, the ancient Greek father of horticulture, said, 'To be a successful farmer one must know the nature of soil,' and I was delighted to follow his advice.

The Tunbridge Wells block is still preserved, in a deep glass case. It shows a top layer of slowly decomposing plant debris, labelled as fibrous peat. Just below this is a layer of topsoil enriched by decomposed organic matter. The next layer of topsoil below that is nutrient-poor, bleached by the leaching of iron hydroxides into the layer below, which is rich and moist, full of the rainwater that stripped the minerals from the level above. And so it continues, becoming rockier the further down you go. If this column of soil were still in the ground, the top layer would be full of microbes breaking down the organic matter, composting it and making nitrogen available to the plant roots. The topsoil would contain pollinators and predators that spend part of their lifecycle in the earth. The layers below store and purify water for plants, and the composition of the whole would provide a stable anchor for the roots that allow trees to grow.

Of course, soil composition varies significantly depending on where you are. Some soil is too shallow or too wet to provide stable anchorage for trees, but if it's too sandy or doesn't get enough rainfall, it may be too dry to support plant growth. There needs to be a decent thickness of topsoil, and the level of acidity or alkalinity has to be right. There are eighteen different nutrients a growing plant needs. Of these, the plant gets three – hydrogen, carbon and oxygen – from the atmosphere via photosynthesis through its leaves. It gets the other fifteen from the soil. The primary ones are nitrogen, phosphorus and potassium (hence the NPK label on commercial fertilizer bags, 'K' standing for potassium), but other elements such as sulphur, iron, copper and calcium are also important in small quantities, and vary from place to place. Minerals and clay particles bind with water and nutrient particles which are sucked up by the plant, and can impact the flavour of fruit, leaves or grain grown in the soil.

Soil, plus climate and the physical, topographical characteristics of the land give us the concept of *terroir*, the idea of 'land taste' first developed by French wine-makers who noticed differences in the characteristics of wine made from the same grapes in different regions, different vineyards, or even different sections of the same vineyard. Given what I now know about soil composition, this comes as no surprise at all – it's obvious that place will have a profound effect on the fruit grown there. So it's curious and frustrating that the application of *terroir* seems to be owned completely by the wine industry.

I'll confess to having a chip on my shoulder about this. Writing extensively about beer and cider, I've seen – and tasted – the direct effects of *terroir* on hops, barley and apples, but have at times been met with scorn and ridicule when I've spoken about it. Surely if soil, climate and geology affect the character of one crop, they affect the character of all crops? Why would grapes be unique? Looking at the EMR soil display, it's more obvious than ever. And then I realize that it's not a question of whether *terroir* exists for these other crops; it's whether the effect of it can be tasted in the finished product. Wine is sophisticated and complex, romantic and mysterious. Beer and cider are crude, industrial and simple – or so many people mistakenly believe. And from the other side of the fence, wine is poncey and pretentious, whereas beer and cider are straightforward and democratic. If we start talking about concepts like *terroir*, we're in danger of our heads disappearing up our collective arses. When Adrian Barlow explains clearly and succinctly why the world's most popular dessert apples come from New Zealand but taste better when they're grown in England, he's talking about *terroir* but, to the best of my knowledge he never uses the word. I think we should use it as widely as possible, to help educate people that there's much more to the character and quality of the food and drink they're buying than they perhaps realize.

⁂

Fascinated by what I saw at the EMR stand at Fruit Focus, I ask if I can come back and talk to them in more detail about their work, and they very kindly say yes.

The first work at East Malling after its formation in 1913 was to gather, analyse and catalogue all the rootstocks used in the UK and across Europe. Growers had been using specific rootstocks for at least a thousand years by that point, the most popular of which were known by name, but no one had ever attempted a comprehensive collection of them before. EMR's aim was to work out what was being used and how well different stocks performed under a variety of conditions.

The timing was fortuitous. Not only was there the burgeoning interest in the potential for scientific research to aid fruit growing that M. J. R. Dunstan had picked up on, but also scientists like Ronald Fisher, J. B. S. Haldane and William Bateson had rediscovered Gregor Mendel's work and were busy establishing the science of genetics. The scientists at EMR analysed the relationship between the anatomy of a rootstock and its vigour. Geneticists then helped to establish the fundamental rules of inheritance and helped create breeding programmes. This work meant that the behaviour of seedlings could be predicted, rather than having to wait and see how they turned out. In 1921, EMR released a categorized range of rootstocks labelled M1 to M16, and followed this up in 1924 with M17 to M24. The Malling (M) series and later the Malling/Merton (MM) series,* allowed growers to predict accurately the vigour (how quickly and how big it would grow) and precocity (how quickly it would blossom and bear fruit) of their trees. This would eventually allow orchardists to move away from big mature trees that yielded no more than four to six tonnes per hectare, to dwarf stocks that produced smaller, more bush-like trees and yielded 40 to 60 tonnes. The dwarfing M9 rootstock is now the base of an estimated 95 per cent of eating-apple trees in the UK, and 90 per cent of the same in the US and South Africa. Around 85 per cent of all the world's apples are grown on EMR's M series rootstocks. The historic impact of EMR's work in this and other areas is estimated to have contributed £8.9 billion to the global economy.

Different stocks suit different purposes. If you want an ornamental apple tree in a pot, you might choose the M27, an extreme dwarfing stock; you'd pick the less dwarfing M9 if you're planting an intensive trellis of commercial trees; or the M25 for a full-size tree. Full-size, wild trees might look better and are hardier in poor or wet soil, thanks to their bigger root systems, but as well as their higher yield, dwarf trees can be picked much more easily. Whatever you need from your rootstock, you could graft the

* Developed jointly with the John Innes Centre in Norwich specifically to be resistant to the woolly aphid – a problem for rootstocks in warmer parts of world.

same scion wood on to any rootstock and get the same apples, but the size and shape of the tree will be quite different.

In the early 1930s, EMR built a new research facility that was essentially a long trench ten feet deep, with a concrete floor, a roof, and large plate-glass observation windows along its walls. Trees were planted along the side of the trench, so the growth of the roots could be observed. It sounds such a simple idea that anyone could have done it, but this 'root laboratory' was the first of its kind. It allowed scientists to relate leaf and fruit growth to the timing of root growth. The East Malling root laboratory informed advances in soil management, planting density, irrigation, cropping levels, and the competition for resources with other plants such as grasses. It played a fundamental part in the continuing development of East Malling's apple rootstocks, and the establishment of much of today's orchard management practice.

In the 1980s, faced with intensifying competition from foreign fruit imports, the Department for Environment, Food and Rural Affairs (DEFRA) under the Thatcher government thought it would be a good idea to slash funding for horticultural research. Among other cuts, the root laboratory at EMR was closed in 1989 and fell into disrepair.

Twenty-four years later, a grant from the Biotechnology and Biological Sciences Research Council (BBSRC) enabled EMR to repair and upgrade the Root Laboratory, and in 2015 Dr Nikki Harrison took me to see it. A row of apple trees stands on one side of a long, grey, peaked roof about a foot above the ground, a row of grasses running down the other side. We enter via a tiny shed, and go down a flight of steps into something that feels like a *Doctor Who* set. Nikki pulls up the blind covering the nearest observation window, and I'm now looking at the roots of the tree above. They lunge down towards the left of the window, but don't seem as fussed about the right. I start trying to formulate a question about this, but in my head it sounds too stupid to ask.

'We're still developing new rootstocks,' says Nikki. 'Nearly all high-density commercial rootstocks are on M9. It's dwarfing and it's precocious – you get blossom in the second year and a light crop in the third. But it does have some major problems, such as susceptibility to fire blight. It's perfect for dessert fruit because it's easier to pick by hand, but for cider you ideally want something that's got the precocity of M9 but not necessarily the dwarfing. In the UK, especially where the cider orchards are, you get a lot of wet weather, so you want a big tree with good root anchorage, and ideally some resistance to root rot. So there's still a lot of room for improvement.'

The problem is, developing a new rootstock takes time. The first M-series stocks were classified and selected from pre-existing stocks. Developing new ones has always required time, as you have to wait to see how vigorous and precocious they turn out to be, and how disease-resistant they are. MM106 is a popular rootstock for gardens and small orchards because of its versatility – it doesn't have to be staked, and can be left to grow as a freestanding tree or used for fans and espaliers. But like the M9 it's particularly prone to fire blight and collar rot. In 2001 EMR launched MM116, which has all the advantages of MM106 but is more resistant to collar rot, woolly aphid and replant disease. MM116 was first created from a cross between MM106 and M27 in the 1960s. It took over thirty years of trialling and testing before it could be commercially released.

Advances in genetics are now helping EMR to speed up the process. 'The initial identification of rootstock traits was purely by observation and deduction,' says Nikki. 'For example, the more bark there is on a rootstock, the more prone it is to dwarfing. But we can now follow up that observation on a genetic level. Is the connection real? If so, what are the molecular markers? Does it still correlate once it's been grafted? If it does, from there we can work out the specific dwarfing genes. And from that, we can work out how these genes control growth.'

Down here, staring at the soil as Nikki speaks, I'm struck by the contrast between how we refer to rootstocks and the trees and fruit they support. I'm in love with the poetry of apple names, with Wellspur Delicious, Wheeler's Russet and White Winter Calville, with Mabbott's Pearmain, Maggie Sinclair and Maiden's Blush. Apples delight all the senses and we name them in that spirit. But all the rootstock gets is a bureaucratic code that has meant nothing to me until now, and still doesn't do justice to the importance of what it's describing. Before it was classified and selected, M9 was called Paradise and M6 was River's Nonsuch Paradise. But poor old M1 and M3 never even had names. This is the business end of the tree – down in the dark, unloved and under-appreciated.

Thinking of the two partners in a grafted tree, and thinking back to the Greek idea that the scion put its own set of roots down through the wood of the rootstock, I ask

Nikki how scions and rootstock actually work together, how the influence of each is determined.

'Think of an ungrafted tree – the leaves send signals to the roots about what the tree needs. The roots send signals back, forming an information loop that regulates the growth of the tree. In a grafted tree, you have two different trees together. When the top of a tree sends a signal to the bottom, the bottom might perceive it differently, and send a signal back in a different way. Instead of a harmonious loop, you have an interaction between two different genotypes. The DNA of either part doesn't change, but the RNA and the hormones *can* change. We've long been able to observe this, but only now do we have the technology to understand *why* it works with any degree of confidence. We can look at which genes are turned on and off.'

The first plant to have its genome sequenced was *Arabidopsis thaliana*, a small annual herb (or weed, depending on your point of view) more commonly known as thale cress or mouse-ear cress. It has a very small genome (a mere 25,000-plus genes) and has long been used in biological research, but the sequencing still took years. 'Now we've got next-generation sequencing that allows us to do the same with more complex, but similar, plants, so we can infer things from what we already know,' says Nikki. In 2010, a group of Italian scientists sequenced the genome of the Golden Delicious, which has 57,000 genes – at the time, it was the biggest plant genome sequenced to date.

I'm staggered by how far, and how rapidly, the science of plant genetics is moving. We're both staring at the roots through the observation window as we chat. I look at the shape of the roots again. My question now seems more stupid than ever. But I have to ask it. 'Look, I'm really sorry, but I have no scientific background whatsoever, so this is going to sound really dumb. Please don't laugh.' I gesture through the window. 'See how the roots are more concentrated down the left-hand side than the right? Let's say we shoved some fresh water and nutrients down the right-hand side. The tree would . . . know, and it would send new roots down to get them. Wouldn't it?'

Nikki doesn't laugh. 'Yes,' she says.

'So that means the tree must somehow . . . *sense* that there's something good down there, beyond its reach?'

'Yes,' she says again.

'Wow,' I say, relieved, 'that is absolutely *amazing*! How does it know? What does it do?'

'We don't know.'

'You don't *know*?'

'The root senses nutrients and sends signals up the tree. Signals come back down and tell it what to do. We know that part, but we still don't know how this sensing works. It's a major branch of apple research.'

I find myself delighted rather than disappointed by this. One of the big differences between the Greek and Roman pantheons of gods is that where the Greeks had gods and goddesses celebrating concepts such as fertility and nature, the more advanced Romans had a specific goddess representing agriculture and cultivation, reflecting the knowledge that was steadily being accumulated rather than purely natural phenomena.* Pomona was the goddess of fruit trees,

* Such scientific knowledge advanced uncertainly in its relationship with superstition and myth. Stephan at Brighton Permaculture Trust would have been

orchards and abundance, and always carried a pruning hook or spade. She was the goddess who taught humans the principles of grafting and pruning, and in art she was usually depicted with a platter of fruit or a cornucopia. Here was a much more practical goddess: rather than throwing magical apples around to create mayhem, she shows us the tools and techniques to cultivate apples ourselves.

The Romans took Pomona's lessons to heart. At its height, the Roman Empire extended all around the Mediterranean, across most of Europe and well into Asia. Its provinces took in the Silk Routes as well as various different climates. Goods moved freely across it, and new fruits and new horticultural practices were brought back to Rome and incorporated into her gardens and orchards. Pliny the Elder, in his *Natural History*, identified twenty-three different cultivated apple varieties in common use in the first century AD.

We still use the methods Pomona taught us today. But over the past century our knowledge of what we're doing and why has taken unprecedented leaps. Pretty soon, we will no longer have to rely on acting as if we know what we're doing. We've gone from observing that rootstocks with more bark on them are more likely to be dwarfing to being able to identify specific dwarfing genes. The benefits of the work at East Malling have had an effect on every grower and eater of apples in the world. But here, in a simple but science-fiction-like research facility, looking at the

delighted with the work of Marcus Terentius Varro (116–27 BCE), who warned against grafting different cultivars onto the same rootstock. According to soothsayers, this practice 'attracts the lightning and turns it into as many bolts as there are varieties'.

roots of a tree that grow by means we really know nothing about, I have a sneaking admiration for the apple tree and its goddess, and the fundamental secrets they're not yet prepared to share with us.

PART FOUR
Harvesting

September–October

'Lo! sweeten'd with the summer light,
The full-juiced apple, waxing over-mellow,
Drops in a silent autumn night.'

Alfred Lord Tennyson, 'The Lotos-Eaters'

12.

Glastonbury Tor

If Britain really is looking to reconnect with a non-specific sense of spirituality, a feeling of union with the world around us and the rhythms it moves to, there's no better example of this than the Glastonbury Festival.

In recent years, Glastonbury – the largest greenfield music and performing arts festival in the world – has become Glasto or, worse, Glasters, a national event that's a little too large for those, like me, who first encountered it as a well-kept secret – one that was easy to keep because no one else had any interest in discovering it. When I started going in the late 1980s and early 1990s, the only coverage the festival would get would be a hundred-word story on the inside pages of a Sunday newspaper about the number of arrests that had been made on the festival site, and we'd always grow indignant that they never made the comparison between the number of festival-goers and a small town of equivalent size and thus realized how astonishingly low that rate of arrests was. Today, Glastonbury is part of 'the season', and summer hasn't officially begun until the newspaper front pages have featured Kate Moss and friends in little shorts and Hunter wellies.

I haven't been to Glastonbury for years, but friends who still go insist that the special spirit that makes it different from any other festival remains intact. It's the atmosphere that means you can forget all about the bands and still have the best weekend of the year. It's the spirit in the air – or maybe the energy in the ground, thrumming through the ley lines according to every stoned weekend philosopher you have a random conversation with – that makes the festival feel not just like a makeshift city full of like-minded people, but another country, a temporary utopia that runs on alternative principles and, for a few days at least, actually seems to work. You feel it most keenly at the stone circle (which has only been there since 1992, but that doesn't matter), where you used to be able to find Joe Strummer and friends every night, talking with anyone who wanted a chat, and singing and playing, and where hundreds of people cheered each sunrise as heartily as any band on the main stage. It feels different. It feels special. Some of this is due to the ethos behind the festival (and maybe the alcohol and drugs) and the people who run it. But much of it comes from the place itself. *Terroir* doesn't only work for fruit.

Glastonbury sits in a part of Somerset fondly known as the Vale of Avalon, named after the fabled island at the heart of Celtic myth. For many years, Glastonbury Festival's founder, Michael Eavis, would use a few lines from the end of Tennyson's *Morte d'Arthur* (his reworking of Thomas Malory's stories in verse) in the publicity for the event. They come from near the end of the poem, where King Arthur says his farewells to Sir Bedivere. The full section reads:

> I am going a long way
> With these thou seëst – if indeed I go –

(For all my mind is clouded with a doubt)
To the island-valley of Avilion;
Where falls not hail, or rain, or any snow,
Nor ever wind blows loudly; but it lies
Deep-meadow'd happy, fair with orchard-lawns
And bowery hollows crown'd with summer sea,
Where I will heal me of my grievous wound.

In Celtic myth, Avalon is the name given to the Otherworld, the Isle of the Dead, which, in Arthurian legend, became a place of healing and eternal life. The 'Island Valley of Avalon', in Tennyson's eyes, is yet another evocation of the Happy Isles, the Elysian Fields, or Eden. Its 'orchard-lawns' even gave Avalon its name: in Welsh, Ynys Afallon means 'Isle of Apples', from *afal* meaning 'apple'. When Geoffrey of Monmouth wrote his *Historia Regum Britanniae* (*History of the Kings of Britain*), the twelfth-century work which gave us the legend of King Arthur, he called it *Insula Avallonis*. In the later work *Vita Merlini* (*The Life of Merlin*) he referred to it as Insula Pomorum, the 'isle of apple trees':

> The island of apples which men call 'The Fortunate Isle,'* gets its name from the fact that it produces all things of itself; the fields there have no need of the ploughs of the farmers and all cultivation is lacking

* 'The Fortunate Isle' was also the name given to the Canary Islands. That's because Geoffrey of Monmouth lifted this description from St Isidore of Seville (560 to 636 AD), whose Fortunate Isles were yet another evocation of the Happy Isles/Eden ideal, and which were identified as the Canary Islands when the Portuguese began their sea voyages.

> except what nature provides. Of its own accord it produces grain and grapes, and apple trees grow in its woods from the close-clipped grass. The ground of its own accord produces everything instead of merely grass, and people live there a hundred years or more.

Like Eden, Avalon is 'well travelled', having been located in Burgundy, Sicily, other unnamed parts of the Mediterranean and even Australasia. But this part of Somerset has won the prize for the source of the name. Specifically, the Vale of Avalon sits guarded by the Isle of Avalon, now more commonly known as Glastonbury Tor. Festival-goers embrace the site as a magical place. And if they're familiar with Tennyson's claim that here 'falls not hail, or rain, or any snow', they conveniently forget that bit as the festival mud inevitably rises.

It's the Tor that makes Glastonbury magical, the Tor that stands a few miles away from the festival's stone circle and draws every festival-goer's eyes back to it again and again over the long weekend. The ridge it sits on is already high above the Somerset Levels, and then it just takes off and looms up, all on its own. However much you know about geology, it simply doesn't seem natural. And yet it's not man-made. So it's little surprise that it's always been a focus of spirituality, magic and religion. Glastonbury Tor is imbued with so much magic and myth, it's surprising it doesn't still levitate above the land. Legend has it that Joseph of Arimathea brought the boy Jesus here. It's home to Gwyn ap Nudd, Lord of the Underworld, King of the Fairies and leader of the Wild Hunt. Since the Middle

Ages it's been topped by St Michael's Tower, all that remains of a church that was destroyed after the dissolution of the monasteries. The surviving tower simply adds to the magic.*

Of course, while the Tor may look magical, it's hardly an island, you might think, gazing out at the surrounding fields. But those fields were mostly flooded until the monks of Glastonbury, Athelney and Muchelney drained them in the Middle Ages, and they remain above water now – most of the time – thanks to a network of drainage ditches. In Palaeolithic times, farmers lived on the Levels, navigating them via timber trackways, the oldest discovered remnants of which have been dated back to 3800 BC. If the Tor still demands myth-making today, one can only imagine how compelling it was when it rose above misty swamps as if untethered from Earth.

Geoffrey of Monmouth's Arthurian tales were commonly regarded as fact rather than fantasy until at least the sixteenth century, thanks to their stirringly romantic depiction of the birth of a nation. The magical Isle of Avalon lacked for nothing. This version of Eden/Hesperides/Happy Isles was guarded not by nymphs or cherubim with flaming swords, but by three, or possibly nine, maidens, chief of whom was the enchantress Morgan le Fay, or Morgana.

* Glastonbury Tor has this in common with the various peaks known as Mont St Michel or St Michael's Mount across the former Celtic world. High places such as this were formerly dedicated to Lugh, the magical high king who is commemorated on Lughnasadh, the summer festival celebrated on 1 August, halfway between Samhain and Beltane. When Christianity arrived, St Michael, the archangel who defeated Satan in battle, was identified with Lugh and eventually took his place.

This is where King Arthur's sword Excalibur was forged and, later, where Arthur was brought when he was fatally wounded. He lives on there, safe from the ravages of death thanks to the magical power of the apple trees.

Historians and fans exploring the possibility of any factual basis for 'King Arthur' have found several real-world figures on whom he just might have been based, but none of them is a perfect match. One is Lucius Artorius Castus, a Roman career soldier who lived in the late second or early third century. Castus had a distinguished military career in Britain as well as in Central Europe, and commanded a group of Sarmartian horse-riding warriors from the plains of what is now Kazakhstan. He was later stationed at Caerleon, near the border between England and Wales, and also made military expeditions to Gaul. This could all have made Castus' exploits known to the Celts in Britain and northern France who have since staked claim to the Arthurian legend.

The problem for Celts who want to claim the Arthurian myth as their own is that the details – such as we assume them now – don't stack up. Ever since Geoffrey of Monmouth, and especially since Thomas Malory, we've imagined Arthur and the heroes of the Round Table as, literally, knights in shining armour on horseback (or miming with coconut shells). But all through the period when Arthur is supposed to have lived, mounted warriors were rare. The Celts favoured chariots, while the Romans mainly used foot soldiers. Even to an idealistic schoolboy, the legends hit a bum note because we know knights in armour didn't really appear until almost a millennium after the Romans left England.

But third-century Sarmartian warriors *did* ride horses, which were first domesticated on the Kazakh steppe, and they *did* wear chainmail and armour of overlapping scales. If we look at the customs and legends from the homeland of these armour-clad horse warriors, other familiar aspects leap out. Arthur claimed his kingship by drawing Excalibur from the stone. The Scythian god of war was symbolized by a sword thrust into, and then drawn out of, a stone. When Arthur died, he asked for his sword to be thrown in the lake, and asked Bedivere to do this for him. Twice, Bedivere couldn't bear to lose Excalibur and lied to Arthur about having thrown the sword away before finally doing so and witnessing the Lady of the Lake catch the sword. In Scythian tradition, a warrior was tied to his sword, and it was thrown into the sea when he died, and there's a similar story about the friend of a dying warrior lying about having performed the deed. There's even a sacred golden cup in the Central Asian myths that sounds an awful lot like the Holy Grail.

However much truth there is in the Arthurian legends, these key details became part of the story only in accounts written centuries after the events supposedly took place. Maybe there was a King Arthur. Maybe he's just myth. Perhaps he was based on a real historical figure like Lucius Artorius Castus, or maybe not. Whatever the truth, many of the key elements of the story are clearly plagiarized from earlier legends originating in Central Asia, and influenced by the horse-riding knights who first brought them here. Truth or fiction, the legend of King Arthur was born in Kazakhstan, just like the domesticated apple that gave Avalon its name. And now I'm going to pick apples, in a place called Avalon Orchard, on the southern slopes of Glastonbury Tor. I feel like I'm stepping into mythology itself.

❦

We approach Glastonbury Tor on a road that runs along the top of the ridge. As we get closer, the rest of the view falls away, becoming superfluous, and the Tor draws all eyes to it, like the Devil's Tower in *Close Encounters of the Third Kind.*

We park on a residential street at the base of the Tor, and while Liz goes off exploring an antique fair, I follow the path through a gate and out on to the open hillside. There are paths straight up, and others running the circumference of the Tor at different heights. The promised signs pointing to Avalon Orchard, and the apple harvest happening there, are nowhere to be seen.

Out of the tree line, about a third of the way up the Tor, is a path that leads left and right, and I have to make a choice. There are no trees above me, and if there's an orchard on the Tor itself, I'm bound to see it below. I go left, and almost instantly begin to feel that I've gone the wrong way. After 200 yards there are no orchards visible on the lower slopes of the Tor itself, but there is one on a low hillside next to it. I don't see anyone in there. I stop and think about retracing my steps.

'Looking for Avalon Orchard?'

An old hippy sits on the hillside next to a blanket of carved flutes and dream-catchers, blond dreads cascading around his weather-beaten face. I don't think I look any different from the steady procession of tourists climbing the Tor for the view. There's no way he can know that I'm looking for the orchard. But he does.

'It's back the way you came, just on the other side of those trees.'

I thank him and start to retrace my steps.

'It's a nice orchard!' he yells after me.

Avalon Orchard is undoubtedly ancient, but I doubt it's responsible for giving the 'Isle of Apples' its name all those centuries ago. Still, it gives me a thrill to think that these trees might be the descendants of the sacred apple trees that offered immortality to fallen kings. The trees look as if they were once pruned, long ago, but have been left to themselves for many years, with twisted, lichen-encrusted branches curling like the arms of a bad actor trying to cower in fear. The fruit they bear is green and small, like squash balls, tanned red on the side that faces the sun. Sheep graze in the next field, and have clearly been at work here. The cropped grass under the trees adds to the appeal of the place, making it feel manicured and ordered. Sparrows flit and dragonflies buzz among the branches.

We're surprisingly high up: I'd forgotten about the ridge between the valleys we drove along to get here, and having climbed a little further up the Tor's lower slopes, I can see the fields, hedgerows and drainage ditches of the Somerset Levels stretching away for miles.

If you asked someone to name a crop that's associated with Somerset, I'm sure most English people would say apples. In the popular imagination, Somerset is the spiritual home of cider. In our national folk memory, if Big Ben and beefeaters stereotypically stand for London, flat caps and whippets for Yorkshire and kilts and caber-tossing for Scotland, Somerset is the home of the buck-toothed, smock-wearing farm worker with a strand of wheat protruding from a big grin, a floppy hat that does nothing to keep the sun off reddened cheeks, and a costrel of cider in one hand. It's a lazy, possibly cruel stereotype, but some of the photos in the Cider Museum in Hereford show it was once very true. This is some of the best apple-growing country there is, with steep escarpments rising from the Levels, perfect soil and plenty of rain. But being tucked away in the south-west, far away from most of the country's big cities and with relatively poor transport links, producing cider made more economic sense here than growing eating apples. Somerset feels easy-going, relaxed, and driving around its winding lanes, you see handmade signs offering cider at the farm gate around every corner. But Avalon Orchard belongs to the National Trust. The apples from here will be used to make juice.

Avalon Orchard is no longer guarded by enchantresses, but by silver-bearded men in orange National Trust volunteer T-shirts and combat pants. There are about twenty or thirty people here, most of whom regard me coolly as I walk through the gate. I'm not sure how I must look to them, a man on my own, definitely not part of the National Trust branch that organized today's event, just someone who found it buried deep in a website and decided to come down all the way from London to take part. It's 20 September today, a couple of days before the autumnal equinox, and the idea of picking apples on Glastonbury Tor so close to such an event was too enticing to resist.

There's a level terrace at the top of the orchard before it

slopes down into the vale, on which a couple of gazebos have been erected, with informative noticeboards and activities for kids inside. There are a few families gamely taking part, but the kids don't seem too interested in the educational displays that have been laid on for them. They do, however, seem fascinated by the simple fact of apples, growing on trees, out here on the hillside. Every time I see small children around an apple harvest, their excitement is the same. It's a purer version of the euphoria I still feel myself when I see laden, bejewelled trees. The orchard is busy and people are focused on their tasks, but I also feel a low vibration of celebration – this is more than just work; it's an event.

I look for someone in charge and find Hayley Dorrington, a National Trust ranger who covers North Somerset. Like many of the buildings the National Trust also looks after, this place has obviously been through a long period of neglect before someone recognized its value and stepped in to save it. Hayley says the National Trust have been looking after the place for about ten years, and no one has any idea what the apple varieties are, who planted them, or when.

I explain that I'm hoping to take part in different types of apple harvest, and that I want to start with a simple, traditional one on full-size trees. Hayley wastes no time in getting me to work. There are teams of two or three around the orchard, picking apples, filling wooden crates with them, and doing essential maintenance work such as building up the stepped path from the terrace into the orchard and clearing back the most overgrown brambles that threaten to engulf the trees nearest the hedge.

Hayley finds me a tree halfway down the hillside that's still full of fruit. She introduces me to another volunteer and we work as a pair. We set up a light, aluminium ladder with three legs, each individually adjustable. The ground in orchards is always uneven, often rocky, and here it's pockmarked with rabbit holes. But tripod ladders like this ensure a firm footing on any terrain, which is reassuring on such a steep hillside. I place the feet and adjust for the right angle to give me steady balance, initially trying to negotiate around the endless piles of sheep shit, and then simply ignoring them, gazing at the treasure above.

Harvesting of any kind is about a delicate balance of timing. Too early, when the fruit isn't quite ready to fall, and you might be missing the chance of allowing it to get to full ripeness. Leave it too late, and it will fall to the ground and, unless it's a cider apple that doesn't mind a bit of bruising, you've lost it. So you want it when the stem's grip on the branch is just starting to loosen, when it can be plucked with a simple twist and gentle pull. If it resists that, it's not ready. Depending on the variety, some apples are just about ready at this time of year, and some have been harvested a month or so ago, but the heart of the picking season is October.

My new partner holds the base of the ladder while I climb up. This is harvesting at its most basic: picking by hand, twisting and pulling, and passing down small handfuls at a time. The other volunteer carries a cloth basket on straps over his shoulders. The basket has a detachable base, held in place with thin rope tied through loops, so when the basket is full you can place it down gently, unfasten the base and allow the apples to roll out into the crate rather than tipping them in and bruising them. I'm soon mesmerized by the work, breathing deeply, completely absorbed.

In the middle of some clusters there are apples no bigger

than grapes, weaklings who have somehow survived the thinning process. But even here, the apple tantalizes, and I'm reminded of the myth that gave us that word. *The Odyssey* recounts the story of Tantalus, a Greek king and mortal son of Zeus who was invited to dine with the immortals on Olympus. Accounts vary as to what he did to upset them – from eavesdropping on the gods' conversations and telling tales to killing his son and tricking the goddess Demeter into eating his shoulder – but, as punishment, Tantalus was sent to Tartarus in the Underworld, where he was confined to a pool of water overhung with fruit trees. Whenever Tantalus bent to drink, the water receded. Whenever he reached for the apples and other fruits above him, the boughs twisted out of his reach – always tempting, never yielding.

Here, at the top of a three-legged ladder, there are apples that I could just about reach if only I climbed higher than the ladder's fourth step. I've been told that it's dangerous to do so, that this is my limit and to go beyond it would risk the ladder over-balancing. But the fattest, most vivid fruit on the whole tree is half a reach away, again and again, and if I just go that extra step . . . But there's no need, and I don't risk disaster. We fill two wooden crates from the one tree and carry them up to the top of the orchard.

You can buy a small juicing kit from eBay for about £70, and they have one set up here: a small wooden cylinder about eighteen inches high that first chops the apples and then squeezes the juice from them. Everyone gets to have a go at operating the press – though the kids seem a little less excited about it than everyone was hoping they would be – and we all get to taste the juice afterwards. It comes out a rich, reddish-brown, and it's full-bodied and robust, with some good, structured tannin and a nice sharpness. To my palate, this would make it work better as cider rather than juice, but it will be blended with apples from other orchards, bottled and sold in National Trust shops.

I want to go back to do more, but people are starting to pack up – I got here quite late. As the rangers begin putting away their ladders and baskets, I mention that I know about all the legends of the Tor. As it's now owned by the National Trust, what are the actual facts about its history?

'Well, it's on the ley lines . . .' begins Hayley.

'Is that a fact?' asks her boss, raising an eyebrow. That conversation ends abruptly, and after thanking Hayley and her team and swapping contact details, I decide to explore the Tor for myself.

The truth is, it's a freak of nature. What is now Glastonbury Tor once sat within a layer of soft clay and limestone, but that eroded to leave a harder-core deposit of sandstone rising in a smooth dome. The slopes were terraced – no one knows by whom or how long ago – but evidence of human habitation goes back to the Iron Age, and the site always seems to have had significance for people of all religions as well as those with no religion at all. I've walked only a few yards before I come across some kind of ventilation shaft or manhole whose cover has been spray-painted with a complex, mystical pattern of geometrically perfect interlocking circles and petals, against a backdrop that resembles a nebula in deep space.

I get to the top, and the wind tugs at me, as it does the thirty or forty other visitors enjoying the view. We can see to the Severn Estuary to the north, and in all directions are fields of bright green bounded by darker hedgerows, a silver glint here or there indicating the drainage ditches that have kept the Levels above water for most of the last 800 years.

I realize I'm feeling a deep sense of euphoria and well-being. I sit for a bit at the top of the Tor, bathing in it as I take in the sweep of Somerset. The place simply makes you feel this way. Some believe it is earthly energy, others that it is God. I don't think it's anything supernatural, but that doesn't make it any less meaningful or pleasant.

There are certain shapes and ratios that have an effect on us, that suggest significance to us. Church and cathedral architects have always known this – it's part of their plan. Sometimes, nature does this herself. There's no conscious purpose, no agenda, but the effect is the same. When we encounter such creations, we can be forgiven for thinking that magic must have been involved. And that's why Glastonbury Tor, the Isle of Apples, simply had to be central to the Arthurian legend.

Apples are central to Celtic mythology in a much broader sense, too. The myths of these islands are more melancholy than those of the warm Mediterranean, the apples cool silver rather than blazing gold. They take place in a climate of mists, and the apple's dual roles as a gateway to the Otherworld and as a symbol of eternal life become indistinguishable.

It's a short walk from Glastonbury Tor down into the town itself. I've been to the Glastonbury Festival many times, but Worthy Farm is actually closer to the village of Pilton, six miles east, and this is my first visit to Glastonbury proper. Its collection of shops is unlike any I've seen, even among the stalls at the Festival. They offer a hundred different ways to find inner peace: totems, cloaks, masks, scents, potions, carvings and magic wands. Glastonbury is a scrapyard of spiritual belief, where damaged, broken or incomplete people can come and create their own pattern of meaning. Buy an offcut of Buddhism, weld it on to a basic pagan chassis, graft on some occultism, a Native American animal spirit guide, a bit of Gaia and finish with a coat of stoner hippy bullshit, all negotiated through a joss stick-scented miasma, in sight of the magical Tor.

We'll never really know how, or when, the apple completed its migration to Britain from its birthplace in Kazakhstan. We know the Romans were experts at cultivation, but there's no actual evidence of apple cultivation in Britain from Roman times. Here, mythology can help us out. We know that only small, sour crab apples were native to Britain until the descendants of Kazakhstan's apples arrived here. It's hard to imagine those crabs being at the heart of so many stories, possessing so much significance, so the apple must have been established in Britain by the time of the Celts. Mythology reveals what factual history cannot, which is why we cling to old myths, and continue to create new ones.

13.

Dragon and Castle

It's mid October, and in Herefordshire the leaves are finally, slowly, starting to turn. Autumn arrives in stages, and when you pay attention the trees remain green much later than you think. Now they're fringed by a light golden haze, but still bright and full.

At Dragon Orchard in Putley, the crop sharers are gathering. In 2001, Norman and Annie Stanier launched a Community-supported Agriculture scheme, an idea that attempts to build links between farmers and the people who consume their produce. At Dragon Orchard, in return for an annual fee, you get to visit the orchard (which isn't open to the general public) at least once a season. Most people do this on one or more of a number of organized open days that feature guided walks, talks and other events that link into what's happening at that particular time of year. On the four main weekends, the Staniers open their house to the crop sharers, cooking a lunch on the Saturday that's accompanied by a selection of cider and fresh apple juice produced from the orchard. And as the October weekend coincides with harvest time, every crop sharer gets to take away a carload of eating and cooking apples, cases of cider and apple juice and jars of Dragon Orchard jams, jellies and chutneys.

This weekend's activity kicks off with Norman taking us on a guided walk through the orchard. Different apple varieties are harvested at different times, and some sections have already been done. The trees closest to the house are almost denuded of fruit, which is being packed into big wooden crates. Each one is a cornucopia: small, shiny, rose-red apples in one crate, huge green apples the size of grapefruit in the next. When they're picked and packed like this, the sweet, musky, 'appley' smell of ethylene, the hormone that makes apples ripen, is concentrated and assertive, reminding me of nail varnish and fruit compote. There's a visual poetry to the composition of the scene: the dwarf trees, still sprightly and leafy despite losing their fruit, the plain, sturdy crates, and the bright, shining apples piled up inside them, are simply pleasing to the eye. I now understand why, when Bill Bradshaw began photographing apples and orchards, he couldn't stop.

This has been a fantastic year for fruit production. 'The quality is very good, because the weather has been perfect,' says Norman. 'Last year we had the driest September for a hundred years, and the apples fell from the trees without much sugar content in them. It was rainy this year. The winter was cold, but not too long, and September was really stable, with warm, sunny days to ripen the fruit, finishing it off, and cool nights with no heat stress.' With weather like this, when the apples fall, the dew sits with them, cushioning and wrapping them in cool grass, preventing them from rotting.

Norman is confirming what Adrian Barlow from English Apples and Pears told the press at the end of September. This year – 2015 – is set to be the best harvest of eating apples for twenty years, with a forecast of 160,000 tonnes picked, compared to 141,000 last year. There was quite a lot of rain in August, but most of it fell at night, which is a dream for apple growers. Dry days and wet nights that are cool but not too cold mean the crop should be particularly juicy. The only concern now is that the weather stays dry through the picking period.

We walk up a gentle slope, past alternating rows of red and green apples, and then along three rows of Dabinett, rusty crimson baubles weighing down the branches so they hang long and graceful in jewelled layers, making the tree look like it's wearing an Elizabethan ball gown. The channel beneath the trees, kept free of grass and weeds that would otherwise compete for nutrients, is bare earth frosted with lichen, dotted with windfalls. As the Dabinetts are just starting to fall now, this means they'll be ready for harvest in about ten days.

Curving back around and descending the slope, we walk through alternating rows of Ellis Bitters and Browns which were planted in 2002. They look quite similar, but on closer inspection the Browns are smaller and redder while the Ellis Bitters are bigger and pinker. These apples make fantastic cider when blended with Dabinett and the shining, green Michelin, but aren't great on their own.

Finally, we come back to Dorothy's Orchard, and it looks very different from the rest. This is the part I've been waiting for since May, and it doesn't disappoint. Most of the apples here are very late, and they're all still on the trees, a riotous palette of green, gold, yellow and red, the quinces at the end of each row shining like the sun, just as Norman had promised. The Spartan holds my gaze, so deeply purple they're mesmerizing. Liz has accompanied me on this trip. She's stuffing windfalls into her pockets, and I urge her to take a bite of the Spartan.

'What's it taste like?' I ask.

She looks at me apologetically. 'Like . . . an apple. It's really lovely.'

'Can you be any more specific?'

'Well, it's incredibly crisp and light, but it's full of flavour. Sorry, that's the best I can do.'

I wish more than ever that I could dive in myself.

When we get back to the house, the crop sharers are loading up their cars with this year's bounty. Once Upon A Tree, the cider company based at Dragon Orchard, has launched a new range this year under a new brand named Harry Taylor, after Norman's grandfather. They're more commercial, lower in alcohol and less wine-like than the core Once Upon A Tree range. Crooked Branch is the drier variety, while Thrown Hat is medium, 'celebrating the feel of having light, air and space in the tree,' says Norman proudly.

Then there are the fresh, pure apple juices. 'Discovery' is made from the single apple variety of the same name. It feels fuller in the mouth than cider, with all the sugar still there in the body. It has a complex flavour that reminds me again of fruit compote more than simple apples.

'It's harder to make good juice than it is cider,' says Norman. 'You can mess around with cider. With juice it's all about picking the fruit at the right time – down to the day – and then handling it with the utmost care, getting everything right, and by everything I mean hygiene. You've only got the juice to play with.'

The crop sharers mainly tend to be middle-aged couples whose children have grown up and moved away, who have reached or are approaching retirement age. They're here because they like apples, because they care about the countryside and like to feel part of an orchard. But they're also here because they retain an energetic curiosity about the world. One weekend when I visited, Norman had an Indian chef come in and demonstrate how to make a meal. The next time I was here, he took everyone off to visit a local bakery. When the crop sharers aren't here, they're off travelling or exploring somewhere else, and every now and then someone will do a little presentation of something new they've found.

Today, one member wants to tell us about Kolomna Pastila, a project she's discovered in Norfolk. It's an alliance between Drove Orchards in the country, which specializes in old apple varieties, and the old town of Kolomna, a suburb of Moscow. Its focus is the celebration of a long-forgotten Russian sweet made from marshmallowy egg whites and apples. It was a popular delicacy in the nineteenth century, but was swept away by war and revolution and slipped from memory. In the early twenty-first century, Elena Dmitrieva and Natalia Nikitina from Kolomna rediscovered the recipe and began making Pastila again. It proved so successful that they opened a shop, café and historical exhibit under the wonderful name 'The Museum of Forgotten Taste'.

The two women were inspired by Fyodor Dostoevsky, who wrote in his *Writer's Diary* in July 1876, 'Humanity will be renewed in the Orchard, and the Orchard will restore it.' That inspiration has proved infectious. In 2013

Drove Orchards saw a link between the nineteenth-century tastes of Russia and the forgotten apple varieties of England, and used this as the basis of an international project to celebrate the apple orchard across borders, and the passion people have for apples wherever they're grown. The first fruit of Kolomna Pastila and Apple Orchards, a new company registered in England at Christmas 2013, is a batch of Kolomna Pastila made with Bramleys from Drove Orchards. In 2014, they launched the Apple Road project, which aims to bring together people from around the world in their love for apples, apple tastes and stories about the apple.

The spread of the project is a perfect example of the paradox of living in the information age. One of the main motivations for people like Andrew Jamieson, the owner of Drove Orchards and the driving force behind the UK side of the project, is the element of prophecy in Dostoevsky's words. As the Kolomna Pastila and Apple Orchards website says, 'In our digital age, that creates disharmony and even threatens the very biological existence of the human beings, the apple orchard can be a healing remedy and the hope for the future.' And yet it's the digital age that makes much of their work possible. The group is now in touch with Isabella Dalla Ragione of the Archeologia Arborea Foundation in Città di Castello, in Perugia, Italy, who rescues old fruit-tree varieties from extinction and is now hoping to discover the original apple varieties used in Pastila. It's hard to imagine this global network functioning without the digital technology it rails against. But I don't think anyone is suggesting a Luddite approach to technology. We need it. But we also need to stay in touch with our natural state. There needs to be a balance, and it feels to people like Norman and Annie Stanier, the crop sharers and the members of Kolomna Pastila and Apple Orchards that we're currently way off-balance.

After lunch (and a whole box of Pastila), Norman takes us in convoy to Castle Fruit Farm, a few miles away between Dymock and Newent. He describes it as a big commercial farm, which grows cider and dessert apples as well as other fruits.

Chrissie and Michael Bentley used to farm sugar beet and barley, but sold up in 2001. They were about to go on holiday to Seattle when 9/11 persuaded them to go to Tuscany instead. British farming was being devastated by foot-and-mouth disease and Michael was out of ideas, but in Italy he was inspired by the Slow Food movement. With no previous experience, Michael and Chrissie decided to buy a fruit farm, and embarked on a steep learning curve and rapid modernization of the farm.

The 150-acre farm is divided into four main areas: half is devoted to dessert apples, Bramleys and pears. A quarter grows plums, and another quarter grows cider apples. Squeezed into the gaps there's a small plot of cherries.

This is large-scale apple growing. The cider apples go to Bulmer's and most of the dessert fruit, pears and cherries go to supermarkets. But the Bentleys are also passionate believers in sustainable farming and the importance of local economies. They are active members of the Newent branch of the Transition Network, a charitable organization that aims to inspire local communities to be self-sufficient by building networks and shopping local, keeping money in the community, where it can recirculate.

The Bentleys sell their produce at local farmers' markets and at their own farm shop, where Chrissie also sells a range of apple juices, chutneys and sauces made from the orchard's bounty.

Michael, tall, laconic, in his mid fifties, takes us for a walk around the orchard. The apple harvest is three-quarters done, but they're working today so we'll get to see a commercial harvest in action. After picking apples on a tripod ladder on Glastonbury Tor, I want to witness harvest on a commercial scale. I imagined that it would be quite different from my previous experience, assuming that the bigger the operation, the more mechanized it would be. It's true of every aspect of agriculture and production, surely?

For cider apples, it is true. Soon after the apples are picked they're going to be turned into a pulp, so no one cares what they look like or if they get bashed around a little. In fact there are many cider-makers who believe fruit that's been a little worn around the edges makes a better drink. The dew in Dragon Orchard's long grass may keep the trees in perfect condition, but here the grass is shorter and it doesn't matter if cider apples sit and ripen on the ground for a few days. The windfall tells you that the stems on the rest are beginning to weaken their hold on the branch, and on harvest day some of those very specific and expensive-looking pieces of *Thunderbirds*-style machinery I saw at Fruit Focus come out for their day (hopefully) in the sun. First, a mechanical arm with a grabber at the end takes hold of the trunk of the tree and gives it a hard shake. It only takes a second, and 95 per cent of the apples fall in a flash rain of colour. Then along comes what is essentially a giant vacuum cleaner, sweeping the apples into piles and sucking them up into a big hopper. So long as the rain stays off and you're not trying to work in a mud bath, the cider-apple harvest is easy.

But dessert apples need to be visually perfect. They can't be damaged, or they'll be rejected. 'We have to grow what people want, and people buy with their eyes,' says Michael.

British apple growers are at the weather's mercy. Herefordshire gets half the amount of sunlight that the south of France does. The French can get a yield of 60 tonnes per acre, while 40 is more typical here. So there's no room for waste, and the fruit has to be picked very carefully, to a specific standard. That means dessert apples need to be picked by hand, from the branch – even on the biggest commercial farms.

Castle Fruit Farm grows the most popular apple varieties in the UK – Gala, Cox, Egremont Russet, Jonagold and Braeburn, as well as a few smaller varieties such as Smitten, a new Gala/Braeburn cross that seems to be showing a lot of promise here in Herefordshire.

We make an unscheduled stop beside rows of beautiful red apples that we simply can't walk past. Michael wasn't planning to stop here, but the whole group simply halts as one, gazing at the fruit. These are the Braeburns. They've loved the recent dry days and cool, wet nights, and a bumper crop is due to be harvested a few days from now.

The apples are planted on trellises, and one of the features that stops us in our tracks is that, with such a good crop, it's more like looking at a solid wall of fruit than individual trees. This is the result of the work done on rootstocks at East Malling. Six-foot stakes are planted at regular intervals with two rows of wire running between them. Dwarf trees need to be supported, so bamboo canes are tied to the wires between the stakes to create the trellis. The trees are planted just 40 centimetres apart, and pruned severely to send spurs along the trellis rather than outwards. This gives a huge increase in productivity per acre compared to traditional methods – and Michael needed to achieve that if the farm was going to be a viable business – but it also maximizes the fruit's exposure to the precious light. It sounds and looks unnatural in the sense that it's hard to imagine these as trees at all, but it's not without its own kind of beauty. As well as being grown for their fruit, apple trees have long been used for decorative purposes too. They can be trained to fan and espalier, creating pleasing shapes against walls and trellises, and this was highly fashionable in ornamental pleasures gardens such as Versailles. Louis XIV's legendary gardener La Quintinye would probably think he could do a more aesthetically pleasing job than Michael Bentley, but he'd certainly appreciate the technique.

Today is the 'second pick' of Gala apples. The apples on any tree don't all ripen at the same time, and yet they have to be picked to tight specifications of size and colour. So on the first pick you only take those that meet the required standard. If the fruit measures at least 60 mm across, it's picked for the supermarkets. If it's between 55 mm and 60 mm, it's picked and sent to local schools. If it's below 55 mm, it stays on the tree for another ten days, after which you return to see if the extra space, light and energy left by the harvested fruit has allowed the rest to grow sufficiently. This is the problem with the phrase about picking the 'low-hanging fruit' first – in reality the fruit at the bottom is often the last to ripen.

This all contributes to a whopping wage bill that accounts for up to 60 per cent of Michael's total costs. The three most costly parts of apple growing are thinning, harvesting and pruning. Another huge chunk of costs for growers like Michael can be attributed to the grading, packing and marketing of apples that takes place after the harvest, but this is done by growers' associations rather than the growers themselves, who therefore can't do anything to try and cut costs in this area. Labour costs on the farm are always at the forefront of any grower's mind.

Apple picking has historically relied on itinerant workers showing up at harvest time, and people used to travel across the west of England to work on their preferred farms. Obviously wages were an important factor in this, but so was cider. Two to four quarts a day was typical as an important part of the overall remuneration. As well as being enjoyable, cider was a vital source of clean liquid when water couldn't always be relied upon.

But the practice of payment in kind was frowned upon. It was common in more industrial areas for factory owners to pay their workers in tokens rather than cash, redeemable only at the company store for products at grossly inflated prices. The 'Truck Acts' of the nineteenth century outlawed this practice to protect workers from being exploited. But when the Truck Acts were extended to cover agricultural workers in 1887, it didn't go according to plan – farm workers wanted and expected a portion of the pay to be liquid in a different sense from the one intended by the financial community. The Report of the Royal

Commission on Labour said 'they would be very reluctant to accept a cash substitute' and, legal or not, the practice carried on.

The arrival of machines with grabbing arms and giant apple hoovers, not to mention tractors and combine harvesters in other areas of agriculture, meant the market for itinerant labour faded away after the Second World War. But as dessert apples still need to be handpicked, farmers like Michael Bentley now turn to Europe for their labour force. The team at Castle Fruit Farm come each year from the town of Chepelare in the Bulgarian mountains, arriving in July for the plum harvest and staying until the apple harvest is over at the end of October. It's hard work. When the apples are perfectly ripe, they have to be harvested at precisely the right time. Now, at the height of the season, working hours are 7.45 a.m. to 4.30 p.m. seven days a week. The workers live onsite in rudimentary self-catering accommodation converted from old farm buildings. Chepelare is a popular ski resort, so when the harvest is done they go home and work in chalets, bars and shops through the winter.

This is a great example of why the debate over immigration and jobs is not quite as simple as the broad brush arguments from either side would like it to be. The arrangement works perfectly for young people from a Central European ski resort, but wouldn't suit most British people looking for jobs. The work is seasonal and temporary. The Bulgarians are fortunate in that they can move between Herefordshire and home and combine complementary seasonal work in both places. There isn't really an British equivalent, and what British person would want a job that only lasts for four months a year, pays minimum wages and requires them to live onsite for that time?

Eventually we catch up with a picking train, heading steadily down an avenue of Galas. A tractor pulls a trailer that's been converted for the picking team. These trailers are specially made and can be steered, which is essential for navigating the narrow gaps between the densely planted dwarf trellises. Four people stand on each side of the trailer, arrayed on platforms at different heights, and the tractor pulls them very slowly down each row, each working within a specified height band. The apples are placed carefully on big wooden pallets at the back of the trailer, and a loader follows on behind, taking away the full bins.

The grading is done by the simple expedient of a piece of card with a hole in it: if the apple passes through the hole, it stays on the tree. Seeing the whole thing in action, I can see how time-consuming – and therefore expensive – it must be. Michael Bentley is worried about what the future holds.

'I pay my workers at an hourly rate rather than by piecework,' he says, 'and the introduction of the new living wage is going to have a huge impact on anything this labour-intensive. Labour already counts for 60 per cent of my total costs, and it's about to go up by 30 per cent. There are half as many fruit growers in Britain now as there were fifteen years ago. There are going to be far fewer when this goes through.'

It seems harsh to begrudge people working so hard more generous rewards for their labour. But the problem for the grower is that, at the same time as labour costs are increasing, there's severe downward pressure on the price at which they can sell their fruit.

'The biggest apple-growing country in Europe is Poland,'

says Michael. 'Now, traditionally, they send their crop east. But because of the unrest in Ukraine and the EU sanctions against Russia, all those apples are now coming west. We've got a massive glut of apples, which means you have to sell them cheaper.'

'So would it be better for you if we left the EU?' says one crop sharer, a righteous smirk tugging the corners of her mouth.

'I know why you say that, but as part of the EU we enjoy free movement of labour. If we left, I'd lose my workforce,' replies Michael.

There don't seem to be any easy answers for the commercial apple grower.

My mind goes back to East Malling. While I was there I also met Dr Mark Else, who specializes in resource efficiency in crop production. He calls it 'precision horticulture'. 'There's a huge amount of variability within an orchard, with different trees cropping different yields,' he told me, 'So we're developing imaging systems to map the variability within an orchard, bringing together different disciplines. These imaging techniques can recognize and count the fruit on a tree – which is extremely difficult to do by eye – and can use that to forecast the yield and variability across the orchard. It will be able to quantify the difference the size of the tree canopy makes, inform and improve pruning techniques, and map *terroir* characteristics. It'll massively increase the efficiency of orchard planning and harvesting.'

This made me think of Harry Taylor throwing his hat

through the branches at Dragon Orchard to check his pruning technique, now immortalized in the name of Once Upon A Tree's newest cider. I told Mark the story. He smiled and said, 'That's great! But our research would have been able to tell him whether he needed to use the same hat size for each variety, and why he was using a hat in the first place.'

I'm not sure what Harry Taylor would make of that. But I think Michael Bentley will be throwing his hat in the air when he hears about it.

14.

Demanding Perfection

And sometimes, it takes only an instant for everything to come crashing down.

Stocks Farm, near the village of Suckley, perches on the border between Herefordshire and Worcestershire, in a narrow, green valley in the Malvern Hills. It's perfect for apples and hops, and is famous for both. It grows a hundred acres of apples, including Gala, Braeburn and Red Windsor as dessert apples, and Dabinett and Harry Masters cider apples for Magners and Bulmers. On 22 August, Richard and Ali Capper were looking forward to the start of their dessert-apple harvest in about a week's time. They were particularly excited about the Red Windsors. Every commercial farmer grows Gala and Braeburn, but rather than being introduced from New Zealand, Red Windsor was first discovered in Suckley, and has turned out to be sweet and tangy with a robust, complex flavour and an alluring deep red hue. It was an early variety, the first to be picked, and Tesco had contracted the entire crop.

Then, on a day when people around the country were grumbling about 'bloody summer weather', the Cappers saw 'a white wall of hail approaching the house'. The freak summer hailstorm lasted just four minutes and was confined to this one valley. The rest of the region's apple growers were unaffected, but for Stocks Farm and two other farms in the valley, the apple season was over.

Four weeks later, I was at Stocks Farm for their annual hop walk, a celebrated event in the brewing industry. (As well as helping Richard run the farm, Ali Capper is marketing director for the British Hop Association.) As we toured the hop yards, I spotted a mound of apples, a loose, sprawling pyramid fifteen feet high, and went closer to have a look. The apples were beautiful – or should have been. I could immediately tell they weren't cider apples because of their size and colour, their uniformity of shape and elegance. But they were covered in pockmarks up to a centimetre across. Very few of them had broken the skin; they were just shallow indentations. But the space these apples were in – a scrappy bit of concrete, sheltered by a few rough, wooden boards, in a farm that's otherwise beautiful to look at – made it clear that Tesco wouldn't be taking any of these apples for sale.

When I asked Ali about the pile of apples, she filled me in. 'We've never known a hailstorm in August before, let alone one like that,' she said. 'No one's going to buy that as a crop of eating apples now. This kind of thing is covered by the insurance, but of course if we make a claim our premiums will shoot up, so I'm now phoning around cider and juice makers to see what kind price I can get for them.' She stopped, her eyes widening. 'I'm sorry, I really can't talk about this any more.'

As Ali said, a combination of insurance and cut-price fire sale to anyone who wants to press the juice from 900 tonnes of ruined eaters should mean the farm doesn't lose out financially. But that's not the point. Apple growing is a year-round activity. The orchard is a beautiful place to be in, and there's a genuine pride in raising fruit that people are prepared to pay good money for. In four minutes, the Cappers watched a year's work – work that they loved – wiped out.

Knowing the story, I would happily have eaten one of the apples from the pile in the farmyard if I had been able to do so without the risk of asphyxiating. They still sent out all the cues that say 'this is good eating' and I'm certain the flavour wouldn't have been affected in the slightest. But that's not how we operate.

I've no idea who originally said, 'The first bite is with the eye,' but I first heard it from the creative director of an advertising agency I used to work for. We were working on a beer campaign together, and talking about the importance of 'the drinking shot' and how the beer simply had to look like the best beer you could ever taste. Ten years previously this ad agency – like all the other great London agencies who at one point made advertising such a wonderfully creative discipline that they started to ask themselves if it might just be art – used to write press ads with beautifully intricate and engaging copy that seduced and delighted the reader with its wordplay. By the time I was working there, press ads didn't have words on them any more, save for a headline, a logo and a website. We've undoubtedly become more visually led, and in a field of too much choice we make buying decisions in the blink of an eye, often bypassing conscious thought altogether.

While I was at the same agency, I did a project for a

publisher that included a market research exercise in which we put a bunch of crime novels on a table and asked people to choose the one they would most like to buy. As I watched them, I noticed no one picked up a book that had a huge black and yellow tarantula crawling around the spine. Rationally, it was a great bit of design. Commercially, it was suicide. I asked the group members why they wouldn't go near the book.

'I don't like spiders,' said one.

'But . . . it's only a *picture* of a spider,' I said.

'I really don't like spiders,' she replied.

'Look, if you read the blurb, the book isn't about spiders at all. The cover artist is just using it as a symbol for danger. Spiders have nothing to do with it.'

'I don't like spiders and I don't want that book in my hands, OK?'

We process visual information quickly, and it goes deep. Our brains make connections from visual stimuli that we're not even aware of unless someone points them out. So even though I know there's nothing wrong with the pockmarked apple, even though my choice might be irrational, I doubt I'd pick it up in a supermarket, and I might not even notice my prejudice.

We're hardwired to respond to visual cues and, in an over-communicated world, we're also being trained by commerce to respond more strongly to them. If a picture really does paint a thousand words, visuals are the way to communicate strong messages quickly and clearly. Supermarkets are both driving this process and in thrall to it. With regard to apples, it's been happening for over a century, and to understand it – and see the warnings of where it could lead – we need to look to America.

In 1862, Henry David Thoreau wrote that the apple 'migrates with man, like the dog and horse and cow: first, perchance, from Greece to Italy, thence to England, thence to America; and our Western emigrant is still marching steadily toward the setting sun with the seeds of the apple in his pocket, or perhaps a few young trees strapped to his load.' The apple tree is as American as it is British.

When the first New England colonists took sacks of grain, saplings and cuttings with them, almost all their crops failed. Apples – apart from a few nasty crab apples – are not indigenous to the American continent, and imported trees failed instantly. But when they took apple seeds, that old genetic diversity inevitably meant there were some seedlings that took to America just as much as the colonists did. As soon as they were planted and grew, America had its own apple trees, completely unique to the continent. Apples – as fruit, fresh juice, vinegar, jelly, apple pie and, most often, fermented into cider – kept the colonists alive. And as the settlers began moving west, one man helped them ensure a steady supply.

There are numerous films about the life of Johnny Appleseed. Along with Daniel Boone and Davy Crockett, he's one of America's real-life pioneers whose story has been turned into a fairy tale. Between 1885 and 1950 more than 300 books and plays were published about him. In the first half of the nineteenth century, John Chapman, as he was originally christened in 1774 in Leominster, Massachusetts, explored the wild frontier before the settlers got there. Every year, he took apple seeds from cider mills in Pennsylvania and headed west, planting orchards in the wilderness.

In the sanitized Disney version, he's a saintly man who preferred the company of Native Americans and children to the colonists, tramped the wilderness barefoot, clad in sackcloth with a tin pan on his head; and was unfailingly kind to animals. The settlers would arrive in their white-covered wagons, tired and hungry, to find a new Eden waiting for them, full of juicy fruit, and the kids would exclaim, 'Aw, gee, thanks Johnny Appleseed!' before the man himself would emerge from the trees, followed by an entourage of cute raccoons, rabbits and squirrels who would play with the kids as they ate.

In reality, John Chapman was a fearsome evangelical preacher belonging to an obscure Swedish sect, a very shrewd businessman and was obsessively in love with a thirteen-year-old girl. He really did walk barefoot through the wilderness, really grew apple trees and really loved animals and Native Americans. A vegetarian, he refused to ride a horse or chop down a tree, and once punished his foot for squashing a worm by throwing away his shoe. But the apples he grew were unfit for eating. We know this because he planted them from seed. And out of all those new American varieties that emerged, only a tiny handful would have had the balance of sweetness and acidity, with little or no tannin, that we accept as eating fruit. The only thing most of Johnny's apples were good for was fermentation. But that suited the grimy, desperate colonists of the real world just fine. America is a country that was built on cider, and Johnny Appleseed was arguably its architect.

John Chapman's apples were richer in diversity than anything Americans or Europeans had yet known. But less than fifty years later, those who came after him would work hard to undo his miracle.

Within a few years of his death, entrepreneurs were combing his orchards, seeking out promising apples and grafting them and crossing them, looking for the best shape, the best look, the best storage potential and the best crunch. Not everyone approved of the new commercial approach. 'I love better [than the more civilized apple trees] to go through the old orchards of un-grafted apple trees, at whatever season of the year,' wrote Thoreau in *Wild Apples* in 1862. He felt that the domesticated, mass-produced varieties lacked character. He thrilled at trees planted from seed, like John Chapman's, and felt they were purer and more honest than cultivated varieties.

But America opted for domestic crowd-pleasers over Appleseed's wild apples or the native crabs that tasted like the music for the shower scene from *Psycho* sounds. Anything that didn't make the grade was ripped up and discarded. It was a gold rush. There was plenty of space for grafting and propagation, and the perfect apple could feed America and the world. If someone found it – just one tree, growing alone, unnoticed but ideal – they could make millions. Eventually someone did.

Jesse Hiatt found a chance wild seedling on his farm in Madison County, Iowa, around 1872. He tried to chop it down twice because it didn't sit neatly with his rows of planted trees, but each time it came back stronger than before. After ten years, the tree produced its first fruit. It was a beautiful strawberry colour, streaked with darker red. Hiatt tasted it and said to his wife that this was 'the best apple in the whole world!'. He named it Hawkeye, and in 1893 he submitted it to a fruit show in Missouri organized by Stark Nurseries. Prizes were offered for high-quality, heavy-producing fruits. Hawkeye, with its unusually elongated body tapering to five points at the base and its distinctive colouring, caught the attention of Clarence M. Stark, President of Stark Nurseries, who tasted it and exclaimed, 'My, that's delicious – and that's the name for it!'

But fate was still toying with Hawkeye and Jesse Hiatt. Somehow, the label had fallen off the box containing his apples. No one knew what this miraculous apple was called, or where it came from. Stark decided to repeat the show the following year, hoping that the mystery apple would be submitted again. It was, and this time its label was intact. Stark bought the tree and the rights, and spent $750,000 launching the renamed Red Delicious across the United States. By the time of the Second World War, it was the most popular apple in the country.

As Thoreau noted, the apple travelled west with the pioneers, spreading across the continent, until it found its new Eden in Washington State. Up in the Pacific Northwest during the height of the growing season, Washington's Yakima Valley gets an hour more sunlight than California, giving the fruit grown here more time to grow. Here in the high desert there are far fewer pests and bugs than in other growing regions. The volcanoes keep the soil deliciously fertile. The only problem is, it's a desert: only an inch of that scant rainfall hits the ground during the growing season. The problem was solved around a century ago, when the valley was irrigated with a network of canals that channelled the meltwater from the winter snowfall on the peaks. Yakima became America's fruit bowl, at the centre of Washington State's 65,000 acres of apples, which now account for over 60 per cent of the entire US crop.

Bill Bradshaw and I spent a couple of days in Yakima at the Tieton Cider Works, staying at Harmony Orchards with the owners, Craig and Sharon Campbell. Craig's grandfather moved to Tieton in the 1920s, not long after the valley was irrigated, and planted the orchard's first apple trees. Craig's family have been tending the orchards ever since.

Harmony Orchards grew exclusively eating apples until 2008, when Craig, who has a genuine fascination for apple varieties, was introduced to cider apples for the first time and planted a test block of cider varieties. Tieton makes excellent cider, and it's becoming a bigger part of their business. When Craig takes us on a tour of his orchards, we see varieties like Kingston Black, which are squash-ball-sized in Britain, the size of tennis balls here. The colours seem brighter, and I swear some varieties are actually glowing with an inner light.

Here on the valley floor we can't see the mountains, but you can tell where they are thanks to the clouds they've pulled around themselves like shrouds. Craig drives us on long, straight roads, past mile after mile of orchards, with the tall concrete cubes of controlled-atmosphere storage facilities every half-mile or so, each one indicating an apple-packing plant.

The journalist Carey McWilliams published a book

about migratory labour in America's farms in the late 1930s. He called it *Factories in the Field*, and wrote what remains a perfect description of an apple-packing facility. Inside the plant we visit, Hispanic women (they depend on itinerant labour here too) in blue overalls line both sides of a wide silver conveyor. First the apples are scrubbed, washed, dried and then waxed. The women take out any with blemishes, and it seems each one has her own brief on what she's looking for. Apples that are left by one woman are picked up by the next. Each woman has a pile of boxes by her side, with soft blue polystyrene liners to cradle each apple individually. They're sorting for colour and size, looking for apples that are identical to the last one they picked out. By the time a box of these apples arrives at the supermarket, they'll look eerily identical, separated from the rest of the crop that looks a little different.

In the UK, this kind of work is done by growers' associations. In the years since I was in Yakima, places like this have become even more sophisticated, with scanners that can now analyse the inside of the apple without even puncturing its skin. Several people tell me that I must visit a British apple-processing facility, but when I make various requests to do so, I'm ignored. I don't think there's anything going on in there that I would be alarmed by but, at the business end of apple production and retail, most people are curiously reluctant to talk. Perhaps they think I'll be horrified by how scientific it all is, or maybe they're just too busy to talk to writers. Either way, the Americans have no such qualms, and so I gain an understanding of what happens to apples after the harvest on a far bigger scale than anything in the UK.

While I'm in Yakima, I also get to handle a Red Delicious for the first time. I'm fascinated by it, because apples have been written about so much more in the United States than they have in Britain, and in the literature the Red Delicious is notorious. Over the course of the twentieth century it became less of an apple variety and more of a brand, and was continually bred to look bigger, shinier and redder. Now I have one in my hand, I'm shocked by it. To start with, it's huge – at least six inches tall. It's so deeply purple, so shiny and curvy, it looks simultaneously regal and sleazy. It's desire and temptation itself, lurid and lascivious, the thief of Snow White's innocence.

In the original fairy tale, the Queen's apple is half red, half white. In the Disney film, the apple is transformed by a magic potion. Before the wicked queen dips it in her bubbling cauldron, the apple is unremarkably green. It emerges covered in slime that forms the shape of a skull, and when the slime drips away, it's bigger and redder, obviously America's most famous and alluring apple at that time and ever since. The witch intones:

> Let the Sleeping Death seep through.
> Look on the skin, the symbol of what lies within.
> Now turn red to tempt Snow White
> To make her hunger for a bite . . .

The film-makers most likely alluded to the Red Delicious aesthetic simply because that was the most recognizable apple variety in America at the time. But to its critics, the analogy is inescapable. It's all about the alluring look, and who knows 'what lies within'.

For obvious reasons, I don't get to taste a Red Delicious, but everyone I speak to who has eaten one describes it as floury, mealy, watery and largely tasteless. Breeding primarily for visual appeal has turned America's most famous apple into a husk. And while it's still the market leader in America, and while we still buy with our eyes, recent sales trends suggest that looks aren't everything. In 1999, there were more Red Delicious sold in the United States than all other varieties of apple put together. By 2010, its market share had collapsed from 51 per cent to 31 per cent. Gala, which is pretty but not as beautiful, and has a sweeter and more complex flavour, increased its market share from 8 per cent to 20 per cent over the same period.

Whenever you talk to supermarkets about buying trends – which people usually do when they're unhappy about those trends – the standard reply is that 'this is what the consumer wants', often accompanied by a shrug of the shoulders. The poor supermarkets are helpless to intervene, responding only to the fickle preferences of the happy shopper. I've heard this so often that I've been tempted to channel generations of mums and shout, 'If the consumer wanted you to jump off a cliff, would you do that too?'

I find this particularly irritating because one of my first jobs in advertising, straight out of university, was working on the British launch of Tropicana Pure Premium orange juice. At the time in Britain, we had heavily processed, pasteurized juice made from reconstituted orange juice concentrate that was cheap and tasted awful, and we had freshly squeezed orange juice that was expensive and tasted nice. Tropicana established a third category – 'pure squeezed juice' – that didn't count as freshly squeezed because it had been gently pasteurized to give it a longer shelf life. But it was way better than the cheap stuff because it was made from 100 per cent juice and hadn't been reconstituted from concentrate. It was the best of both worlds. It was kept in the chiller cabinet like freshly squeezed, and tasted very similar to it, but it was significantly cheaper.

So pure squeezed juice was an attractive, interesting product with the potential to get people to spend more money on orange juice, and the big supermarket chains were very interested in that. The trouble was, it required people to learn all about the difference between 100 per cent juice and juice from concentrate. It needed a brand with a decent advertising budget to tell the story. So the supermarkets worked closely with Tropicana, fully cooperating with the brand to help launch it. Once the brand was established and successful, they all launched their own-label versions and cashed in.

In this instance – and almost certainly in the case of countless other product launches – the supermarkets weren't just shrugging their shoulders and saying, 'This is what the consumer wants.' The consumer didn't want pure squeezed orange juice because they had no idea what it was until Tropicana spent a lot of time and money explaining the concept. If they want to, supermarkets can and do influence consumer tastes rather than following them blindly.

I was curious to know whether the drive for visually perfect fruit, the debate around style over substance, was being shaped by the supermarkets, or whether they were simply following consumer demand. What did they really look for in apples? Would they prefer to sell a few big brand varieties or was there scope for something more nuanced, for educating the consumer about apple varieties the way they did with pure squeezed orange juice? I contacted the PR departments of a few supermarket chains, who agreed to put these questions to their fruit buyers.

Greg Sehringer, who buys apples for Waitrose, is the only one who agrees to talk. In Greg's view, if there's one word that sums up what a buyer is looking for, it's quality. But what does that mean? 'It's quality of flavour, appearance, texture, sweetness and level of colour, which are all dependent on the different expectations for each variety,' he says. 'It's the biggest challenge we face as we source through the seasons across the two hemispheres.' Meaning, of course, that if the quality isn't good enough close to home, they'll look further afield, just as the British apple-buyer did back in the late nineteenth century. Happily, the reality of British fruit is very different today, and Greg Sehringer agrees with Adrian Barlow of English Apples and Pears about the quality of British fruit. 'There's no doubt, in my opinion, that thanks to the maritime climate in the UK and the fact that our Braeburn and Gala apples spend longer on the tree than in continental Europe, in vintage years our apples taste exceptional.'

On the questions of fewer varieties and the importance of visual impact, Greg says what I expected him to – that it's all driven by consumer demand. But he backs this up with similar observations to those I made while working in advertising. 'People are busier and spend less time looking at the in-store displays while shopping. Appearance and colour inevitably have an instant impact on which products customers choose. However,' he adds, 'something more important is happening to our marketplace, and that is that the flavours that customers used to select – the tanginess of a Cox or a Braeburn – are being replaced by the sweeter varieties such as Gala, Pink Lady, Jazz and Ambrosia.'

We have more choice than we ever did, and so we rely on visual cues to make quicker decisions. It doesn't mean we think this is more important, just that we have to use it to create a short cut. This makes sense. In my book research with the tarantula-covered paperback, our main question was: if a customer walks into a bookshop (we did it back when there were a lot more bookshops) without a specific title in mind, and there are 45,000 books in stock, how do they navigate that choice down to the one or two they end up buying? We're constantly being told that choice is a good thing, and there's the assumption that if choice is good, the more choice there is, the better. But too much choice means we're engaging less and less with the decisions we make and the things we buy.

And while we assume the consumer wants all this choice, once they've picked one of the options, they want it to remain consistent. 'Customers want the confidence that the Jazz apples they bought this week will eat the same as the Jazz apples they bought last week. They want to know that the fruit will last the desired length of time at home, and that they will not have waste,' says Greg.

This is how we've been trained to buy. Of course supermarkets are complicit in this, either creating or reinforcing our purchasing habits, but we're at fault too. We buy our groceries the same way we consume scurrilous celebrity gossip, complaining about how terrible it all is while actively engaging in the exchange. But we do have a choice: Waitrose and Marks & Spencer stock more apple varieties than any other supermarket, and specialist greengrocers stock even more. When shopping is something you can afford to spend a bit of time on, and you're not just motivated by price and convenience, the option is there to explore and choose and celebrate interesting varieties and ingredients and lift food out of its branded and commoditized rut. The popularity of celebrity chefs on TV and the sale of cookbooks every Christmas suggest we're doing just that. But so far, our shopping habits tell a very different story.

PART FIVE
Celebrating

October

‘Therefore does this apple fall perpendicularly or towards the centre? If matter thus draws matter; it must be proportion of its quantity. Therefore the apple draws the Earth, as well as the Earth draws the apple.’

Isaac Newton in conversation with William Stukeley, 15 April 1726

15.

The National Fruit Collection

This is the week the weather changed.

It happens around mid October and even in the city, if you're paying attention, you can feel the abrupt snap in the air. You leave the house one morning and not only is it cooler, the air smells different, thinner and lighter. The leaves are only just starting to turn, but the season has definitely gone over.

In a couple of weeks we'll be celebrating (or trying to avoid) Halloween. Today's over-commercialized/Satanic/slightly dodgy event (delete according to your prejudices) has its roots in the Celtic festival of Samhain, versions of which have been celebrated across Ireland, Scotland, the Isle of Man, Wales, Cornwall and Brittany since Celtic times. Traditionally, Samhain was celebrated for a full twenty-four hours, from sunset on 31 October to sunset on 1 November. It was timed to fall halfway between the autumn equinox and the winter solstice, and stands opposite Beltane in the wheel of the year. Most of the harvest is in, although some apples are still on the trees. Cattle have been brought down from their summer pastures, and some are being slaughtered for winter. Bonfires were lit for practical reasons, but also for their cleansing powers, just as at Beltane.

And just like Beltane, Samhain was a liminal time, when the boundaries between this and other worlds became permeable. Beltane, in celebrating the return to light, gave us stories of fairy kings and queens dancing in the meadows. But Samhain faces the dark half, and inevitably the Otherworld in this instance conjures up the Underworld and the dead.

The Christian Church initially celebrated All Saints' Day in May, a day to remember the dead, especially the saints who were killed when Christians were still being persecuted. Before that, the Romans had honoured the goddess Pomona in mid August. Both these festivals were moved to coincide with Samhain, and the traditions of all three combined to gives us All Hallows' Eve, or Halloween. Just like Christmas, the name derives from the Christian festival, but many of the customs and arguably even the 'true meaning' predate that.

At Pomona's festival, girls would make their individual marks on apples which they floated in a bowl of water. The boys then had to try to take a bite out of an apple, using only their mouths. Supposedly, when a boy succeeded in taking a bite, he would marry the girl whose mark was on the apple. Another version of apple-bobbing involved an apple being hung from a piece of string, with boys competing to take a bite out of it. The first boy to succeed would be the first to marry. Once again, the apple symbolized desire, romance and loss of innocence.

The apple may have been overshadowed by the pumpkin on Halloween itself, but the spirit of Samhain hangs over the whole month of October in orchards across the northern hemisphere. In North America, families drive out of the cities in long convoys to fruit farms, where they drink freshly pressed apple juice, eat cider doughnuts and choose a pumpkin for Halloween. In the UK, many orchards hold more modest celebrations, but one in particular works on a scale the Americans would recognize.

The UK's National Fruit Collection now lives at Brogdale Farm in Kent. The first attempt to create such a collection was begun by the Royal Horticultural Society in the early nineteenth century, with the main aim being to establish a catalogue of fruit and get rid of anomalies such as one plum cultivar that had 120 different names depending on where it was grown. The first catalogue of the National Fruit Collection, published in 1826, listed 1,400 apple cultivars. But then the fortunes of the Royal Horticultural Society went into decline, and when they recovered, fruit studies were not as high on the agenda.

By the mid-nineteenth century, the British apple industry was being killed by imports. Something had to change, and change came in the form of Robert Hogg. Hogg accepted the post of secretary to the RHS's fruit committee in 1860. Not only was he Britain's leading pomologist, he also had the energy, drive and vision to counter the threat to British apples from foreign imports. He published his *Fruit Manual* the same year, which went on to be revised and expanded until it contained information on the origins and characteristics of all the main cultivated varieties, along with advice on the conditions that suited their growth. According to one fan, it made 'the history of fruit as interesting as a fairy tale'.

Hogg's close involvement with the activities of the Woolhope Club and the initiatives to revitalize apple cultivation in Herefordshire taught him that there needed to be similar initiatives on a national level. In 1883, Hogg and his colleagues on what was now the Fruit, Vegetable and Herb Committee organized the Apple Congress at the Society's Great Vinery in Chiswick Gardens. The aim was to gather and catalogue all the apple varieties being grown in Britain. A committee of fifty experts from across the country convened to identify, crosscheck and create a definitive list that ultimately included 1,545 varieties. These were put on display to the public, who were given questionnaires designed to evaluate the best varieties in each area and their top sixty dessert and top sixty culinary varieties overall.

The event almost never happened, because the RHS wondered 'whether it was or was not departing from its dignity in allying itself with an exhibition of Apples'. But it was already obvious that 1883 was going to yield a bumper apple crop, and Hogg pushed the idea through to fruition. The Apple Congress effectively gave birth to the modern fruit industry, and resulted in a new momentum to catalogue and preserve traditional English varieties. In 1922 the RHS and the Ministry of Agriculture began the National Fruit Trials at Wisley in Surrey, which aimed to identify the best cultivars for commercial production. Concurrent with this, a new fruit collection was started. After the Second World War, the collection outgrew the space at Wisley, and cuttings were taken from every variety to propagate onto new rootstock at Brogdale. By 1960, the process was complete.

In 1989, the same Thatcher government that decided the

fruit industry would do just fine if it slashed horticultural research at East Malling decided we could also do without the National Fruit Trials at Brogdale, and the programme to help the British industry identify promising new commercial varieties came to an end, effectively leading the British apple market to rely on European imports for years. But as a result of public pressure, the National Fruit Collection was preserved. Today, the 150 acres of Brogdale Farm are home to 2,200 different apple varieties, 520 pears, 320 plums, 320 cherries, 200 red, black, white and pink currants, and 150 gooseberries. It's now run by the University of Reading's Farm Advisory Services Team (FAST), which gives advice to Britain's 150-plus fruit growers on all aspects of growing, and conducts trials and experiments here to hone and develop that advice.

The chequered history of the National Fruit Collection suggests long periods of establishment indifference confronted by the passion of a few individuals and a level of enthusiastic interest among the general population, and it seems that pattern still holds today. I meet Tim Biddlecombe, CEO of FAST, who repeatedly tells me how lucky he feels to have been able to spend a career in this industry. But he's also frustrated at the level of support he gets for the work done at Brogdale. FAST had to pitch to take over the collection, and it won because it realized commercial concerns had to take priority. But that doesn't mean the organization agrees this is how it should be. 'Our productivity is going down now, compared to the rest of the world,' says Tim. 'Scientists used to get funding that allowed them to research areas they felt might hold the biggest benefit. Now, all the funding is project-based. Finish the project, and you're out of a job. The Italians – who we met recently – aren't under this kind of pressure, so they're free to think creatively. We're a commercial organization and we have to do research that reflects growers' needs and disseminate our findings to the growers that pay for them. The UK doesn't even really have a pomologist.'

Later, when we're talking about genetic diversity, I ask if Brogdale has incorporated any new varieties from the astonishing range of wild apples in Kazakhstan. Tim admits he's never visited the country. While the Americans are propagating the past and possible future of the apple using the genetic material from the Tien Shan mountains, the British can't afford the plane fare, given the current state of funding the industry receives.

The short-sightedness of people and institutions who could be funding research that is necessary to build a competitive apple industry – let alone broader concerns about building our horticultural understanding – is astonishing. Whether they're on the commercial side, in genetic research or working on heritage and diversity projects like Common Ground, the only reason the apple is grown in Britain today is because a handful of people give a shit about it and work in the face of indifference or even opposition to preserve it.

The buzz at Brogdale's National Apple Festival suggests the public are on the side of the apple enthusiasts. By 11.30 on a Saturday morning, despite the steady, dispiriting drizzle, the event is so busy we're directed to an overspill car park and need to navigate a muddy track back to the centre.

The old farm buildings were converted into offices when the collection was moved here, but there's also now a courtyard of shops selling locally grown farm produce. They're doing brisk business today, and then beyond them, in the Festival space itself, are the same ubiquitous stalls that crop up at country shows like the Bath & West and any kind of food and drink festival, some of them selling vaguely relevant goods such as kitchen knives or chopping blocks, others a random collection of leather goods, knitwear and household knick-knacks, as well as the obligatory hog roast. Brogdale is also home to the Faversham Miniature Railway, which is, apparently, the only nine-inch gauge railway in the UK that's open to the public. It's incredibly popular among the visitors, probably because the little steam engine is faster, cheaper, more comfortable and more reliable than anything belonging to Southeastern trains. Parents smile blissfully, using their kids as an excuse to experience a train that works.

There's plenty more to keep the kids entertained: falconry exhibitions, wellie-throwing competitions, and an apple pie-eating competition that echoes Samhain's apple-bobbing. Curiously, only dads with kids present seem to want to take part in this. They line up with their hands behind their backs and prepare to go face down into their pies, only for the competition to be temporarily halted when a worm is spotted beneath the table. Mayhem almost ensues when a feral child leaps up and tries to stomp the creature to death, but the woman running the competition, who is for some reason dressed as a scarecrow, gets angry with him and he backs off.

There's much to do here – lots of tea and coffee to drink, and lots of cake and pork baps to eat, so I'm delighted to see that most people who have paid their admission fee have flocked to the main barn to try to get a look at the apple display. Brogdale boasts that this is the biggest apple exhibition in Britain, and I haven't seen anything to contradict that. Over 400 apples from the collection are displayed in ranks of small bowls with neat little labels giving the variety's name, place of origin and the name of the person who 'raised' or discovered it. Slices of each type of apple are laid out in bowls to sample. You can buy a plastic bag for £3 (or two for £5) and fill it with your favourite apple varieties.

The size of the queue is enough to make anyone question the assertion that the British public is only after a few uniform, sweet brands. Of course, this is what you'd hope for if you were organizing an event celebrating the apple, but it's still an arresting and cheering sight. I hadn't realized how fascinated people are by apples. They taste in thoughtful silence, steadily filling their bags as they go along. There's even lots of polite English jostling, body-checking, sharp elbows and little side steps around the most intriguing varieties.

The display includes some varieties that are new to me: fat, lumpy King of Tompkins County, striped, regal Lewis's Incomparable, golf-ball-sized High View Pippin and grapefruit-sized Grandmere. Granny Smith is striking, a deeper green, darker hued than its neighbours. You can see its allure, even though the half-dozen examples on display here are more varied than any you'd see in the shops. The Galas are yellow and pink, reminding me of rhubarb and custard chews. And I love that there's a golden, slightly blushing variety named after Gene Pitney.

Mike Austen, an apple historian and orchardist who has worked at Brogdale since 1992, gives a talk on the origin of the apple in a greenhouse full of two-year-old apple trees for sale. Brogdale specializes in supplying older varieties that are no longer commercially grown, like the almost extinct Bossom that Peter May told me about at Stanmer. More recently, they've been advising heritage bodies and 'wealthy individuals' about how to recreate orchards in historic homes, identifying what was grown in orchards at a certain time period and trying to re-establish them today. Ultimately, when there's a proven demand, Brogdale sells scion wood from these forgotten cultivars to commercial tree nurseries, who then start the process of reintroducing them into the food chain.

Mike tells the story of Kazakhstan, the Persians, Greeks and Romans, and when he comes up to date, he doesn't stop. He tells us that the DNA from a Cox's Orange Pippin is not an exact match with the wild apples of the Tien Shan, but it's close enough to remove any doubt about the theory Vavilov developed by observation alone.

He then moves on to explain how a new variety is developed and tested. There is a middle way between grafting and planting a random seedling and hoping for the best. I've often wondered about this when people tell me this apple is descending from that one, or this is a cross between those two. Using manual pollination, you can't guarantee anything the way you can with a graft, but instead of letting an apple blossom be pollinated naturally, you can choose the other variety that pollinates it. Like breeding dogs or horses, the progeny will still be different from their parents and from each other, but you can create a fairly good chance that some of them might inherit favourable characteristics from each parent.

To do this, you choose two varieties that have strong characteristics you'd like to combine. You take all the petals off an apple blossom and apply pollen from one variety directly onto the stamen of another, using a paintbrush. You then put a bag over the stamen to prevent any further pollination by insects. You let the resulting apples grow, and the seeds inside these apples are the children of the two parents.

Mike tells us that East Malling plants 20,000 seeds a year from manually pollinated apples. The seedlings that result are covered in greenfly to test their resistance. Those that survive are grafted into vigorous rootstocks to speed things up, planted out in a field and tested again, sprayed with spores from the main diseases that affect apple trees. Some will hopefully have natural resistance to these diseases, but at each stage fewer survive to go forward. Once a seedling has been identified as resilient, after three or four years it's evaluated for yield. Of the seedlings that are resilient and bear decent fruit, another 60 to 70 per cent are discarded when they're evaluated for the appearance, size and flavour of the fruit. The winners are then trialled on a commercial scale, and any that work are named and released. The whole process takes twenty to twenty-five years, and from the initial 20,000 seeds, an average of six new commercial apple cultivars will emerge. You then just have to hope that the public likes at least one of them.

FAST conducts these trials here at East Malling. Tim Biddlecombe tells me that fruit growers sometimes start conducting their own trials and then wish they'd never started. It costs £30,000 to plant a hectare of trees. Trials to prove commercial viability can quickly become very expensive, and then Brogdale will take over. One current trial has thirteen varieties that have passed the first hurdles and must now be tested for commercial viability. One plot here consists of two rows of each. 'Unless you've got something really exciting, it can be very difficult to break through commercially,' says Tim. 'People are currently buying Jazz, Pink Lady, a bit of Gala and a bit of Golden Delicious. If you're going to break through, you've got to convince people that your new variety looks, eats and stores better than those superstar varieties, and can be grown at a good enough volume and price.'

That doesn't mean it's impossible, or that there are no gaps in the market. Tim points to a tree bearing round, shiny green fruit. 'That's Smeralda. Golden Delicious and Granny Smith are still popular, but they're imported. If you grow Granny Smith in Britain, they don't pick until November. An English green apple with the same look and eating qualities would be good.'

Of the current crop, Tim predicts three may have some commercial merit. But it must be tempting for an ambitious grower. Everyone involved in apple-growing knows the stories of the lottery-win discoveries of Red and Golden Delicious. There's always a slight chance that the next one may be lurking somewhere in the orchard, and that it might be as stubborn and lucky as Red Delicious was.

I wander past the pop-up vintage tearoom, the model railway and the falconry exhibition to get onto one of the orchard tours of the National Fruit Collection.

'How long does the tour take?' I ask one of the women sitting at a desk under a gazebo blocking the entry to the orchard, channelling those who want a tour on foot and those who want to ride on a trailer attached to a tractor in the right direction.

'About an hour, but it depends who you get,' she replies with a wry smile.

We get Mike – a different Mike – who used to work for a growers' cooperative and has been a guide at Brogdale for seventeen years. Mike sees no point in saying much until we're among the trees, and we walk briskly into a relatively new area of the Collection where the trees – two of each variety – are all planted on M9 rootstock. This orchard is five years old. I've seen enough orchards now to spot that it's not commercial. Despite the dwarf rootstock, the trees are spaced more widely than they could be, and are pruned differently. These trees are to be studied, not harvested intensively, and one of the main roles of the collection is as a source of graft wood for anyone who wants to propagate a rare variety. In a commercial orchard, that first-year growth would be cut away. Here it's left to spindle out from the tree, a product for sale just like the fruit. I realize I'm in a living museum, a gene bank to help preserve the genetic diversity of fruit crops.

'This is Britain's contribution to global genetic resources,' says Mike. 'There are going to be dramatic changes in the world's food supply in the coming years. We're getting elements of the Mediterranean climate here in Kent now, and a notable flattening out of seasons.' This is another reason why Kent is a good place for the collection. In 2003 the highest-ever recorded temperature in Britain – 38.3 degrees Celsius – was recorded here.

Another trial in a different field is seeking to evaluate the effects of climate change. For this trial, twenty-five different varieties have been selected from the collection to reflect as much genetic diversity as possible: early flowering and late flowering, those that respond to a high level of winter chill and those that prefer it to be lower, and so on. These varieties have been planted in three different blocks and covered to shield them from the natural weather. The first block will be kept at the outside ambient temperature, the second, one to two degrees warmer, and the third, two to three degrees warmer. All the rainfall will be collected, and one block will receive the natural level of rainfall poured back onto it, with one drier and one wetter. It sounds to me like a vital piece of research, one that could have huge benefits for the future of British apple-growing in an increasingly uncertain climate. So it's dispiriting to learn that even though DEFRA asked FAST specifically to conduct the trial, there was no money available to do so until a legacy to the Trust enabled them to go ahead.

After forty-five minutes we've walked 150 yards and looked at two different trees. There's so much to see, so much to talk about, and for everyone on the tour apart from me, so much goodness to taste. Mike checks the time and speeds up, and we go through the hedgerow to an older part of the collection, with bigger trees planted on more vigorous stocks. They're still pruned quite low but have thick trunks planted two metres apart, and look like trees, rather than bushes.

Running out of time, Mike is desperate to show and taste as many different apples as he can: Decio, which was named after a Roman commander who came north to fight the Huns; small, green, russet Great Expectations; and Junami, which was raised 1,300 years ago in China, and arrived here via Japan.

Finally, Mike brings us to Holstein: 'Very close to being my favourite.' It's bright gold and green and it shines, and when he hands me a piece I put it in my mouth instead of discreetly dropping it on the floor like I did with the rest. Its taste is a great balance of floral sweetness and a nice, sharp acidity. This piece was only flesh, with no skin, and I convince myself that it's going to be OK. When it comes, my reaction isn't as bad as it was to the Kingston Black a few years ago, but soon my throat is like sandpaper, and pins and needles are rippling down my tongue.

Later I meet Dr Richard Harrison, a geneticist at East Malling, who expands on some of the issues Mike discussed on the tour. It's obvious that Brogdale and East Malling work closely together, so it's handy that they're just a few miles apart. Kent may no longer be the 'Garden of England' as Henry VIII once dubbed it, but it's certainly Britain's centre of fruit science.

Richard is optimistic about the future of British apples, but believes there's no time for complacency. 'The UK

apple industry failed to innovate at key points in its past,' he says. 'Production and competitiveness fell, but thanks to a lot of hard work, they've now been turned around. Dessert apple production in Britain has doubled in the last ten years, and grown rapidly in the last three to four years.'

I ask about the concentration of new varieties, and the loss of the old.

Richard is pragmatic. 'Look, my favourite apple is Cox's Orange Pippin. I love it, and so do a proportion of consumers. But it's an ageing variety. Younger age groups want sweeter, juicier apples and crispness rather than a floury texture. At the same time, we want to maximize health benefits such as antioxidants. We're looking at disease resistance to fungal pathogens, but you never get everything you want in the same cultivar. The answer is to bring traits in from the wild, combining them with the desirable traits from domesticated varieties such as Braeburn, Cox and Gala.'

I think of the potential unlocked by the rediscovery of the apple forests of the Tien Shan, and the collection at Brogdale that has nothing from there in it. Should we be doing more to preserve the apple's genetic diversity?

'The genetic diversity isn't being lost! Breeders and geneticists are hoarders – they hang onto everything. In breeding programmes, it's all captured, even if it's not there on the supermarket shelves. Heritage orchards have a tiny segment of the genetic diversity, from a scientific point of view. We have lots of varieties at Brogdale, but the problem is they're not future-proof. But new research on the genome has transformed what we can do. Until a few years ago, we could tell where something was, but not what it did. Now we can locate a specific gene and say what it does and doesn't do. So we go into them and look for pest and disease resistance, flavour, chill resistance – all in a working climate, calculating the mechanism that determines when it buds and so on.'

The apple is different things to different people. To some of us it's romance, poetry, religion, art, history and identity. These are all the marks of a civilized society, so we naturally feel that something is wrong if they're being destroyed or neglected.

But to others, the apple is a livelihood, and you can't expect a farmer to plant and tend an orchard if it's losing money. To make money, it has to compete in an oversupplied market, and to do that, it has to maximize yield and minimize damage from pests and diseases. And to do *that*, you need to have scientists exploring its attributes using the very latest technology they can. And that brings us to the subject of genetic modification.

In 1989, researchers at Cornell University used a 'gene gun' to successfully transfer an antibacterial gene from a Cecropia moth to an apple tree that was susceptible to fire blight. This enabled the tree to develop its own fire blight resistance, and that resistance was passed on through buds and grafts. Genetically modifying the apple tree's DNA massively improved its health.

But when they hear stories like this, many people, particularly in Europe, throw up their hands in horror. Writing that last paragraph, I struggled to keep my language neutral because the image in my head was like something out of David Cronenberg's *The Fly*. I don't want moth bits in my apples. Forget my allergy, if I ate one of those things I might start sprouting feathered antennae and head-butting bright lights.

I'm certain that such concerns are irrational. But enough of us share them that there are currently no fresh GM products allowed in the EU. Scientists like Richard Harrison believe this means we're being left behind.

'I understand the concern with GM,' he says, 'But I think it arises from two different schools of thought. First, there's this perfectly natural reaction that says, we're working with a lovely, fresh product, let's not contaminate it. But second, I think the whole "Frankenstein Food" thing was chiefly down to GM being seen to benefit the company rather than the consumer. My personal opinion is that we should be investing heavily in GM, but that it should be happening in the public sector, for the public good.

'What we're talking about is a set of tools, and those tools are improving. Think of a TV. You wouldn't ban all television just because you don't like one programme. You'd campaign for better programmes. And right now GM is HDTV, and we're still being forced to watch black and white. The stuff they were doing with GM in the 1990s was like silent movies. We need to move to a more product-based rather than tech-based evaluation.

'As of the last three or four years, we have the ability to go and modify a single pair of DNA in any organism. You can edit a single faulty gene. In the US, this counts as non-GM because you're not introducing any foreign DNA like your moth example. All you're doing is changing the existing DNA in exactly the same way a random mutation would. Now we understand the genome, we could make a few changes to it and give an apple complete resistance to disease. You're controlling the fate of your crops, but you're not changing their genetic make-up.'

At the time of writing, Richard Harrison is waiting for the outcome of an EU deliberation over whether CRISPR technology* – the switching on and off of the genes he described – counts as GM or not. If it doesn't, he can start using it at East Malling. I'm amazed how much the technology and science of apple cultivation has moved on even since I started my research. 'Everything's changed in the last three or four years,' says every scientist I speak to at East Malling.

What I find comforting is that the answers to the apple's problems are in a union of the past and the future. The American scientists who visited the Tien Shan mountains in the 1990s estimated that the apples that had been domesticated in the West contained as little as 20 per cent of the genetic material available in the wild apple forests. Those wild apples, now being cultivated in collections around the world (even if we can't afford to take part), could potentially give us five times as much genetic raw material to play with as we ended up with after we focused on a handful of intensively cultivated varieties.

Meanwhile, the technology of the future, the rapid acceleration of our understanding of the apple's genetic make-up, is the equivalent of learning a new language within the last five years that allows us to communicate with apples and change them at the most fundamental level.

* CRISPR stands for Clustered Regularly Interspaced Short Palindromic Repeats. No, me neither.

I'm still not sure how I feel about the GM/Non-GM technology Richard describes, but one thing he and his colleague Dr Mark Else tell me sets an agenda for the debate that puts our fears and ambitions into perspective: based on current global population projections, we need to double the world's food supply by 2050. To put that another way, we need to produce as much food between 2000 AD and 2050 AD as we did between 1500 AD and 2000 AD. And we need to do that at a time when scarce resources such as clean water in underground aquifers and phosphates in the soil are being rapidly depleted.

Spending time at the business end of the modern apple industry is a sobering experience. Food scarcity has been a problem for most of human existence, and the apple has always been at the heart of images of cornucopia and bounty. The English apple industry is now trying to take account of scarcity while dealing with a glut of over-supply from Europe. It's a strange scenario that makes me long to return to the simple delight of the traditional orchard. I'm very lucky that I can.

16.

Bramley's Seedling

As a traveller, I hate getting caught out by local quirks and customs, especially when I'm still in my own country. Abroad, you're tolerated if you don't know the correct pronunciation of a word, but few Brits can resist a smirk when we hear Americans saying *Ed-in-bor-row* or *Worse-ses-ter-shire*. If you commit similar crimes as a Brit, it feels as if the penalty should be worse. So I feel genuine shame and embarrassment after boarding the number 100 bus in Nottingham city centre.

'A day return to Southwell, please,' I say to the driver.

'Suth'll? That'll be five pound. Cheers, duck.'

Of course. With those long vowels, *Southwell* just sounds a little too uncomfortably posh for Nottinghamshire. *Suth-ll* is a no-nonsense local collusion to accommodate the place, so everyone can be comfortable with it in their midst.

After an hour on the top deck of a green bus winding along the rolling English roads of north Nottinghamshire, I arrive in a village, or maybe a small town, where the houses have big gardens with pretty walls and hedgerows, and the shops look as if they've been transplanted from Lewes or Islington.

Southwell is an odd place to find in this part of Nottinghamshire. It's visibly, achingly affluent, in an area that's struggled since the decline of the coalfields. The leaflet from the Tourist Information centre says the town is often referred to as 'the jewel in Nottinghamshire's crown'. One of Southwell's chief claims to fame is that King Charles I spent his last night of freedom here in the King's Head pub (which still stands today, although it's now the Saracen's Head) before surrendering to the Scottish Covenanter army in May 1646. The centre of town is dominated by Southwell Minster, a giant, rather severe cathedral that dates back to the twelfth century, but looks eerily well preserved. It's truly impressive, as well as being a laughably over-sized church for a small town of fewer than 7,000 souls. It has on occasion given the town aspirations to be recognized as a city – a demand that has always been rejected. It turns out that the people who actually live in Southwell pronounce it *Southwell* after all.

It's another drizzly October Saturday, and the pavements and pathways are carpeted with damp, golden leaves. For the first time this autumn, my breath mists the air. This is, of course, how it should be, but I miss the crisp, sunny days of the orchards a few weeks ago, when I was picking apples or watching them being picked.

Church Street, leading past the cathedral and out of the centre of town, is lined with grand villas with names like the Old Rectory and the Coach House. Further away from the Minster, the houses become more modest, until you hit a row of cottages with front doors that open straight on to the busy pavement.

Number 73 has a plaque bearing the name 'Bramley Tree Cottage'. The gardens of this cottage have been reduced, but next door, the number 75 is carved onto a weathered wooden apple, and a Heritage Lottery Fund blue plaque above the house number tells us that:

> THE BRAMLEY APPLE TREE was grown from a pip by a young lady, Mary Anne Brailsford, between 1809 & 1815. It was thought it came from an apple grown on a tree at the bottom of her garden (now No.75). One seedling produced very fine apples in 1837 when the new occupier was Mr Matthew Bramley. A local gardener, Henry Merryweather, later obtained permission to take cuttings from the tree and it was duly registered as the Bramley Seedling.

It's a modest house, a private dwelling with UPVC windows defending it against the traffic noise. There's a driveway down the side with a silver Volkswagen parked in it, the redbrick house on one side, a tall redbrick wall on the other, and a new-build cottage at the bottom, taking advantage of excess garden space. It strikes me how ordinary this place is, how unremarkably typical it is of any house in any street in Britain. And then I realize that's the point: the Bramley apple originated in the garden behind this house not because there was anything special about it, but because a superstar apple variety can spring up anywhere, at random. This particular permutation is responsible for bringing hundreds of people to Southwell every October, some from as far away as Japan, where the Bramley is particularly revered.

It's just gone noon, so after reading the plaque and looking down the driveway and being stuck for something else to do on the site of the Bramley apple's birth, I follow a group of apple tourists into the Hearty Goodfellow, the pub next door. Inside, there's a small apple on every table. As I sit down with my pint, Lisa, the publican, brings me a complimentary apple and cinnamon fritter on a small plate. 'Happy Apple!' she beams, as she goes off to serve similar gifts to the party of about a dozen English and Italian visitors I followed in here.

I decide to test my allergy: I know I can't eat fresh apples, but I can drink cider and fresh apple juice. Where do cooked apples fit in?

I love what happens to the texture of the apple when it's deep-fried. It retains some kind of structural integrity – just – but melts as soon as it's in the mouth. There's sweetness, obviously, but also a gentle yet insistent acidity that begs to be paired with something soft and creamy. It finishes pleasingly rounded, the tingling sweetness lightly tempered by the cinnamon. This is an extraordinarily good apple fritter. If that's what the Bramley is all about, its global fame makes perfect sense. And I experience no allergic reaction at all.

While there are thousands of different varieties of apple – when they're cultivated they're referred to as cultivars – but from an eating and drinking point of view we break them down into three broad types: cookers, eaters and cider apples. This classification is determined by the characteristics of the apple flesh. An apple has three key flavour dimensions: sweetness, acidity and tannin. In the broadest terms, whichever of these is most dominant determines what we use the apple for: we eat the sweet ones, bake with the sharp ones, and make cider with the tannic ones.

In the trade, cooking apples are referred to as culinary apples, which is fair enough. But I've always been confused by the description of eaters as dessert apples. Put the words 'apple' and 'dessert' together and I think immediately of apple pie, and therefore of cooking or culinary apples. And I've never met anyone who thinks an eating apple on its own makes a perfect dessert, nor have I ever seen an apple on a menu as dessert. So I always have to correct myself from thinking of dessert apples as cookers.

The root of my confusion is historical. Today, the eating apple may be a snack for the busy person on the go, or something you have with a packed lunch. But from the ancient Persians all the way through to Victorian England, the best banquets in the most opulent dining rooms ended with a dessert course consisting of the finest fruits attainable, as well as jellies, syrups, sweetmeats and candied or crystallized slices made from them. The dessert table was the final flourish, the symbol of the wealth and sophistication that were necessary to source such an array.

But until the Victorian era, the dessert course was separate from the pudding course. We now use the two words interchangeably, with people arguing over which is the correct one. But in the age of great banquets, the cooked pudding always came first, followed by the delicate finish of the fruit-based dessert.

Apples were common in both. They looked great on display on the dessert table with their rainbow of colours. Firm and polished, they gleamed in the candlelight. These prized, named varieties were cultivated for sweetness and crispness, so only the finest would feature. But cooking softens and sweetens the apple, and lesser varieties – apples whose sharp flesh might have appealed to Thoreau in his winter forest but to few others – have been used in the kitchen since the Middle Ages. They added excitement to cereal porridges, bulked out meat dishes and were used in custards, pancakes and fritters. By the seventeenth century, they were also being baked with sugar in pastry casings to make apple tarts, pies and puddings, appealing to the English sweet tooth (and benefiting from the cheap price of sugar from the West Indies). English cookery writers began to argue that the apple pie or pudding was morally superior to the extravagance and complexity of French cooking. These simple, homely dishes were not just tastier, but more patriotically English. Even the French seemed to agree. A Monsieur Misson, visiting England in the 1690s, wrote:

> Blessed be he that invented the Pudding, for it is a Manna that hits the Palates of all Sorts of People; a Manna better than that of the Wilderness, because the People are never weary of it . . . To come in pudding-time, is as much as to say, to come in the most lucky Moment in the World.

While the pudding course never really went out of fashion, as the cultivation and selection of an ever-increasing number of apple varieties became a medium for displaying both wealth and taste, the dessert course that followed remained the focus of apple appreciation. By the late nineteenth century, discerning diners discussed the flavour of their apples as passionately as they would the finest wines. Eventually, this aesthetic approach would have to bend to the trend of commercialization coming out of the United States, and apple growers would have to start concerning themselves with such bothersome matters as yield and pest resistance if they wanted to stay in business. But for a while this uniquely British approach elevated apple varieties to their greatest ever heights of fame. And right into the middle of this lofty air of appreciation came a unique culinary variety.

Some of the details on the plaque at number 75 aren't quite correct, which is unfortunate when you've cast them in metal. Mary Ann Brailsford did indeed plant some pips in the garden, which she took from apples her mother, Elizabeth, was using to make an apple pie. Mary Ann married and moved away a few years later, so the tree would have been tended by her mother, along with the rest of the cottage garden, until she died in 1837. Matthew Bramley, a local butcher, didn't buy the cottage that year, but in 1846.

In 1856, Henry Merryweather was just seventeen years old when he spotted some very fine apples being carried in a basket by a work colleague. Although young, Merryweather worked in the Southwell nursery for his father and had a good knowledge of apples. He enquired where they came from, and when his colleague replied that he'd picked them from Matthew Bramley's garden, Merryweather approached Bramley and asked if he could take cuttings from the tree for grafts. Bramley simply waved him on, inviting him to help himself.

Merryweather built up a collection of trees, the fruit of which he named Bramley's Wonder and Star of Southwell before settling on Bramley's Seedling when he first showed the apple to the Royal Horticultural Society's fruit committee in December 1876. In 1833, he entered it into Robert Hogg's National Apple Congress in Chiswick.

The Bramley's Seedling, then only known around Nottingham, couldn't have hoped to top the list of favoured national varieties at the Congress, but it did receive the highly prestigious RHS First Class Certificate, which was awarded only to the very best plants of any kind.

The Apple Congress stunned visitors with the variety on display in a country dominated by foreign imports. The stars of the show, such as the then-new Cox's Orange Pippin and Worcester Pearmain in dessert and the Bramley in culinary, went on to become celebrated and widely cultivated varieties that remain popular today. Bramley's Seedling was planted intensively in Kent, and swiftly became popular with commercial growers. By the 1940s, it was such a big seller at London's leading fruit and vegetable market that it was known as the 'King of Covent Garden'.

Britain is the only country in the world that has a culinary apple as a major market variety. The Bramley is grown extensively across the UK, and in Northern Ireland the Armagh Bramley, slightly sharper than the English variety, has its own EU Protected Geographical Indication (PGI) status.

Michael Bentley at Castle Fruit Farm in Herefordshire told me that he'd stopped growing Bramleys because no one uses them for cooking any more. That may be why the English Apples and Pears campaign is making the Bramley its main focus. They dominated the stand at Fruit Focus in East Malling, and the Bramley apple website offers endorsements from chefs including Mark Hix, Marcus Wareing and Angela Hartnett. We may not be making apple pies so much (even though the 'traditional Bramley apple pie filling' has just been granted EU PGI status) but the website has recipes for plenty of other puddings (some of which are desserts) as well as main courses, starters and snacks.

There's certainly enough interest in the Bramley to bring people to Southwell from thousands of miles away. The local paper announces that this year's Bramley Apple Festival – the 22nd annual event – is being opened today by Professor Matsumoto, an apple researcher who has flown almost 8,000 miles to be here as part of a delegation seeking to build links with Southwell and capitalize on the popularity of the Bramley in Japan. Roger Merryweather, festival chairman and great-grandson of Henry Merryweather, is delighted to welcome him.

Here in the Hearty Goodfellow, one of the English visitors at the table across the pub is explaining to the Italians the principle of genetic randomness in apples, and the significance of the original tree of any cultivated variety. When Lisa serves them their apple fritters, they ask about the Bramley tree next door.

'Oh, she's still there,' replies Lisa. 'You can still see her. The lady who owned the house died last year and she used to show people around, but it's a bit difficult now. But the tree is still there. She's beautiful, and she still bears fruit.'

After I finish my pint, I walk down to the bottom of the pub garden. And finally, here comes my reaction to the apple fritter. My tongue is fizzy and swollen. Not by much, but I feel like I've swapped my tongue for Jamie Oliver's. No more delicious apple fritters, then.

The high redbrick wall separates the pub's beer garden

from the plot where Mary Anne Brailsford planted her pips. When the wall runs out, further back from the road, a dense line of trees takes over. The parts of the garden I can see look overgrown, and I'm not quite sure where to look. I jump up and down a bit, looking for something that might be a 200-year-old apple tree. Any fruit will have been picked by now, which makes it a little more difficult. I convince myself that a low, bent tree must be it, and later, when I find a picture, I tell myself I definitely saw it. But I really must stop jumping up and down in a muddy pub garden, clearly visible from the road, and move on.*

Back in the centre of Southwell, the Bramley Apple Festival is in full swing. Most of the shops have created apple-themed displays in their windows, and the pubs and restaurants are serving Bramley-themed dishes. It's not all about the apple, but the apple is the inspiration for a broader celebration of food, drink and tradition. Morris dancers wearing white shirts and trousers and straw boaters covered in corn dollies move from pub to pub, playing and dancing to what anyone who grew up in Britain during the 1970s and 1980s would recognize as the Country Life Butter Men song.

In the churchyard of the Minster, the leaves are flame-coloured and in full fall – a steady, delicate flutter. The rain comes on properly now, and of course I've left my umbrella in the pub. Along with several hundred others, I immediately take a keen interest in the food and drink festival happening inside the Minster. The staff on the door

* In July 2016, as this book was going to press, the BBC ran a news story that the original Bramley tree planted by Mary Ann Brailsford is now dying of a fungal infection. Nancy Harrison, who lived at number 76, tended the tree for most of her life until she died recently.

can barely suppress their grins as they struggle to take our admission fees.

Inside, the story of Jesus throwing the moneylenders out of the temple briefly flashes across my memory, triggered by the sight of a comically large cathedral absolutely crammed with market stalls and people shopping among them. Once again I'm struck by the variety of people here. Some are clearly interested in learning more about the different apple varieties on show and are keen to try the produce of local butchers, bakers and chutney-makers. Others – usually in family groups – wander among the stalls with flat expressions and a passive shuffle. It's something to do, something a bit different, a hope that it might keep the kids quiet for an hour or two on a rainy Saturday afternoon in October.

Outside the Minster, I wait for the bus back to Nottingham. There's a proper queue in single file – an extinct phenomenon back home in London – and I suspect this may have something to do with me being the only member of the queue who is (a) not a woman aged over sixty-five; and (b) over 5 feet 3 inches tall.

The driver arrives, and he's a bus driver from some children's TV programme. He can't possibly exist in the real world. He seems genuinely happy to see us, his face beaming beneath his comb-over. He welcomes each of the ladies aboard as if they were his own grandmother. 'That's right, on you come! Ha! Ha! Oh, yes! Come on, that's it, there's plenty of room upstairs! Ha! Ha! Ha!'

I'm the last one on.

'All right, pal,' he says as he stamps my ticket in the machine. 'It's a grand life, in't it?'

It certainly is, pal.

17.

Not an Imaginary Place

I'm more pleased than words can describe to be going to Much Marcle. Until I started my exploration of cider and apples, I actually thought it was a made-up place name, one of those fictional places like Dingley Dell or Ambridge that represents a shared ideal of the English country village. I was delighted to discover that it's real.

Much Marcle is the largest of the seven parishes of Marcle Ridge, the other six being Aylton (real), Little Marcle (real), Munsley (real), Pixley (real – honestly), Putley (real), Little Plinkington (just made it up) and Woolhope (real). Much Marcle radiates out from a busy crossroads where there's a pub, a village shop and a garage. You can't really call this the centre of Much Marcle though, because it doesn't quite live up to such a billing.

Walking along what appears to be the main drag in the bottom of the valley, I'm quite disappointed that it's not the quaint Cotswold stone village of my imagination – it is a real place after all, full of unlovely, sensible bungalows that look like open prisons for elderly tax evaders.

The village hall is also a relatively new building, all tall windows and neat parquet floors. The Much Marcle Brownies have set up in the main hall, serving tea and a range of homemade cakes. The labels on one or two of them gauchely admit to being Mary Berry recipes, but I suspect the main authors of this collection look down on the *Great British Bake Off* and all who watch it as arriviste pretenders to a great and noble tradition.

In the smaller room next door there's a sale of local produce and second-hand books. The one item I would love to buy isn't for sale: a framed print of a map of Much Marcle drawn in 1797. Streets and buildings don't hold much interest for the cartographer: it's all about the plots of land, and what they're used for. Different shades denote hops, orchards, arable and grass, and each plot has a name like Wheelers Knowl, The Stockings or Boyless Meadow, names that could inspire stories or even novels. Some survive today as street names for the open prison bungalows.

The lady behind the stall asks us if we're local, and when we say we aren't she carries on telling us anyway that the bowling-ball-sized onion was grown by Tom, and that these wonderful specimens here are Phillip's tomatoes. There are a great many apples, mainly cookers and early eaters – each handful, each bushel, someone's pride and joy. These are big, honest apples, pockmarked and ruddy thanks to the reality of Much Marcle, not the shiny, uniform fantasy of the supermarket. Our new friend explains that these are good for apple cake, and those are excellent for chutney. All are apologetically cheap.

I'm reminded of harvest festival, which I think I last participated in when I was about nine years old. We had a table at school and we were all required to take something in that represented nature's bounty, or rather, God's gifts to us. I think we were too young to understand the importance of the ritual either way, and I'm sure I remember some kids bringing in tinned goods, which, I now understand, completely missed the point. When you see a pile of freshly harvested apples, it can send you giddy with happiness. And when it's a bad harvest – 2012 was one of the worst in living memory – livelihoods can be ruined.

So it makes sense that we have events like the Big Apple Harvestime celebrations in Much Marcle. I once felt imprisoned by places like this, and surely still would if I was confined to a life of village church halls, as I was as a child. But now I see the goodness of it, and I envy it. It's calm and soothing here among the jams, chutneys and marmalades, the lemon drizzles and Victoria sponges.

This weekend in mid October is the busiest of the year for the Big Apple, the celebration of apples and cider organized by Jackie Denman and friends. This year, events are happening across eight official venues focused around Much Marcle, with a tractor and trailer transporting people between the main ones. There are apple teas, and opportunities to have a go at pressing apples and pears on a traditional wooden press, learn about edible hedgerows and the importance of pest control, even to listen to a concert of songs and poetry inspired by orchards, apples and cider.

We decide to head to Hellens, the most historic building in the village and one of the oldest dwellings in England. It is the centre of the weekend's action, although neither of the words 'centre' or 'action' quite belong in Much Marcle. Hellens is a manor house with roots going back over

900 years, and takes its name from Walter de Helyon, who moved here in the fourteenth century and whose descendants still live in the house. Although parts of it are older and others newer, the whole place has a seventeenth-century feel to it, a dense but elegant cluster of arches, steep gables, lead-panelled windows and weathervanes, set amid courtyards, pathways, secret gardens, waterfalls and barns that today play host to tutored tastings, perry-making and, inevitably, morris dancing. The tearoom sells hot drinks and slices of cake so vast they could sink battleships if anyone could devise a means of airlifting them from here to the sea.

Most of the interesting stuff is in the main barn, around a cluster of stalls that, in line with the rest of the associations Much Marcle provokes, conjure up long-forgotten memories of childhood church jumble sales. The only difference is that where I expect to see old winter coats, worn cricket bats and Airfix models with half the parts missing, there are apples. Just apples. Lots of apples, all different, accompanied by dishes of puree for you to taste and compare.

Further down one side of the barn is the Marcher Apple Network (or Rhwydwaith Afalau'r Gororau if you prefer), a society committed to the revival of old apple and pear varieties from the local area. It was founded in 1993, and if their large display is any indication, they've had a productive couple of decades. Beneath the proud but less than catchy slogan, 'Finding and conserving the old apple varieties of the Welsh Marches', there's information about museum orchards and displays of rescued apple varieties sporting wonderfully arcane names such as Onibury

Pippin, Newton Wonder, Pig yr wydd and Winter Quoining. There's also a wanted list of disappeared and almost forgotten names, missing but not yet given up on.

On the table in front of the display boards sit examples of apples that have been rescued – trophy fruit of burgundy, scarlet, golden, khaki and peach. Behind the table, three wise sages, as old as time, sit straight-backed and inscrutable like mountain-top monks. As a matronly lady delivers them a china pot of tea and slices of cake the size of house bricks, a pretty woman in Hunter wellies brings them an apple from her garden and asks the wisest, oldest-looking panel member if he can identify it. Suddenly everything goes a bit *Antiques Roadshow*. The apple is held gently, turned and stroked, as are long grey beards. Eventually the fruity Buddha produces a penknife and slices the apple open. He scrutinizes the seed pattern, makes some deep, ruminative noises, cuts a slice and eats it. Then he opens a copy of a *Pomona*, a beautifully illustrated book of apple varieties. He doesn't seem to know what this one is.

As a spectator, I'm uncertain. I can't figure out whether I'm supposed to be excited – *Hey, this might be a new variety!* – or disdainful – *You're not very good at this, are you?* So I get bored, and wander off to look at some more cider apple varieties, lingering over each neatly handwritten label:

Bridstow Wasp
Barcelona Pearmain
Gipsy King
Sugarloaf Pippin
King's Acre Bountiful

William Crump
Maiden's Blush

These words are as mystically English as the shipping forecast. I want to chant them, a comforting spell as the nights draw in. Get the order and inflection right, and they will probably summon up cosy, mashing teapots and slices of warm, buttered Soreen out of thin air.

I'm drawn to the *Pomona*, the book the Marcher Apple Network is using to try to identify apples. Copies of the *Welsh Marches Pomona* are on sale at a bookstall on the next table. These books are incredibly beautifully illustrated, with life-size colour plates showing the whole ripe apple, a cross-section through it, and an example of the blossom. Even now that we have high-definition colour photography, the illustrations in the *Pomonas* (*Pomonae*?) are still hand-painted. The apples strangely seem more real this way, vivid and ready to eat, and the combination of apple and blossom together is a visual treat to be drunk in greedily.

When I first started studying apples, I decided I'd look on AbeBooks, the website that aggregates supply and demand for second-hand books worldwide, to see if I could find a copy of Robert Hogg's and Dr Henry Graves Bull's *Herefordshire Pomona*, which they put together and published between 1878 and 1884 at the behest of the Woolhope Club. I often hunt out old books in this way when I'm writing a new one. Usually, a century-old book that's difficult to find but would be fascinating for my project is languishing on the bottom shelf of a dusty bookshop in Gloucester or Washington DC and I can buy it and have it delivered for a few pounds. And sure enough, I found various copies of the *Herefordshire Pomona*, most of them in old bookshops in Herefordshire. Depending on what condition they were in, they were on sale for between £5,000 and £13,000.

Drawings and paintings have been used in catalogues of apple varieties since the eighteenth century. But their approach was selective and there was no overall system of classification. Part of the task the Royal Horticultural Society set itself from its foundation was to clear up the confusion over classification, and that was partly achieved by producing the *Pomona*. But when you look at a copy, you have to ask whether they needed to be quite so lavish just for the purposes of classification and identification. And of course the answer is no. There was something more going on.

The fashion for paintings of plants had been growing through the seventeenth century. The typical table-top still-life composition of fruit inevitably contained apples, and in terms of representing the object, painting of fruit arguably reached its peak in the *Pomona*. The bookstall here at Hellens is selling a CD-Rom – already a dramatically outdated piece of technology, which seems curiously appropriate with respect to Hogg and Bull's *Herefordshire Pomona*. I buy it, and when I finally locate my CD drive, I discover it contains only scans of the colour plates, not the text. Proof, if any were needed, that people bought and still buy the *Pomona* just to gaze at the pictures.

But while the *Pomona* was being produced, still-life painting was coming to be regarded as the lowest of artistic genres. Paul Cézanne proclaimed, 'With an apple I will astonish Paris,' and he did so when he unveiled his *Still Life with Apples* in 1898 which reinvented the form. Cézanne was drawn to apples by their beautiful colours, the variation between them, and their simplicity and completeness of form. The apple symbolizes both Eve and Venus, and was often included in Renaissance-era depictions of both women. 'Painting from nature is not copying the object,' he wrote, 'it is realizing one's sensations.'

From religious art to still life and impressionism all the way through to surrealism, the apple has been a receptacle for any sort of meaning we want to imbue it with, from being a symbol of desire, lust, envy, purity or wholesomeness, to a figure that challenges us to consider how we perceive the world and what art is for. The coloured plates in a *Pomona* don't mean anything except 'this is what this particular variety of apple looks like'. But if we really thought that was true, we'd be calling them something like 'apple catalogues' rather than naming them after a goddess who lived in the woods, tending her trees and revealing their transcendent beauty, before sharing (most of) her secrets with us. These beautiful paintings feel like the least we can do in return.

18.

Apple Day

Saturday dawns bright and clean. Well, I *say* dawns . . . after an evening of cider, perry and Calvados with Bill, his wife, Lisa, and my wife, Liz, I'm oblivious to how the day dawns. But by the time we emerge, crusted and bleary, the sky is as pure as freshly washed laundry and the shadows cast across the Somerset Levels by the low sun have crisp, sharp edges.

Five minutes' drive from Bill's house is a curious hill with one big tree at the top. Just below the summit stands Pass Vale Farm, home of Burrow Hill Cider and the Somerset Cider Brandy Company. We park by the tree, and cross the road to the farm.

By 12.30 p.m., there are around a hundred people in the farmyard. A band sits in one corner playing Dixieland jazz, while an ancient man sits in a chair next to them, hunched over his walking stick, nodding off. Nearby, the apple stand is stacked with all the products of Burrow Hill and its orchards: cider, perry, brandy, an aperitif called Pomona, and a big drum of mulled cider with a hissing flame underneath. Every minute or two, a two-litre jug of cider with mulling spices swirling in it comes out of the drum. The woman serving behind the stall advises us to go for the mulled cider with a shot of apple brandy. After one sip, Liz will drink nothing else all afternoon.

A few feet away, a group called the Woodmen have brought their ancient travelling cider press and are giving a demonstration. On their six-foot-long wooden cart they build layers of apple pulp embedded in straw which is folded in on itself to cover the pulp. This is cider-making at its most traditional. When the juice had been pressed from them, people used to use these packed bricks of straw and apple pulp – known as cheeses – as fuel during the winter.

More than anything else, our eyes are drawn to the big piles of apples on one side of the farmyard. One pile is of a single variety, Kingston Black, gleaming in the sun. Another pile of mixed apples – some green, some blush – consists of around forty varieties. Children see them and just run at them, jumping on top and rolling over them, climbing up lopsidedly, arms wobbling. For a second this strikes me as an extraordinary reaction. But they're only acting on an impulse I think we adults also feel: the sight of so much fruit, all sitting there in a ripe pile, stirs something deep and does something to the brain. The screams and yelps and giddy laughter remind me of what it's like when we see the first snowfall of winter.

'I love what it does to kids,' says Bill, reading my mind. 'They become more childish. Boys start throwing apples at each other.' Every few seconds, a Kingston Black, dislodged from the pile by the frolicking children, rolls by my feet.

I don't want to insult anyone who lives in Somerset, but given that we're in a small farm in the middle of the countryside, I'm surprised by the number of spectacularly tall and beautiful young women among the crowd. They're all wearing sunglasses and interesting combinations of wellies, fleeces, legwarmers and light summer dresses. 'That's Alice Temperley's crowd,' says Bill.

Pass Vale Farm is owned by Julian Temperley. His daughter Alice is one of Britain's most celebrated fashion designers. Recently hailed as the 'English Ralph Lauren', she's the designer of choice for Kate and Pippa Middleton. She lives in London, but has brought friends down for the weekend who are now mingling with the ruddy farmers, women in tweeds and shellacked hair, old boys in red neckerchiefs and hole-filled jumpers covered in stains, and leather-clad hairies. The atmosphere is one of deliciously laid-back celebration. But at the same time, it feels like a society event.

If it is, it looks like the most curious society you've ever seen. Everyone who is anyone in Somerset – and West London – seems to be here. An old man with a long white beard with no moustache wears a burgundy trilby. Another wears a shirt and tie, what might be a Hare Krishna robe and a large fedora. He starts to dance to the music. Bill tells me that last year he was up on stage singing along, seemingly unaware, or uncaring, that the microphone in his hand had long ago been switched off. Sometimes the dividing line between London fashion and Somerset eccentricity seems a little blurred.

But it's not just Julian Temperley's daughter who brings these two worlds together. Our host now wanders across the yard. On the biggest social event of his year, he looks as if he's just rolled out of bed and thrown on some clothes to do a bit of decorating. He wears stained trousers, an untucked shirt and a chunky knit tank top. But appearances deceive, and the casual tone reflects a fiercely independent and headstrong attitude. This is one of the most influential, well-connected and controversial people in the cider industry.

Alex James, bass player for Blur and now cheese-maker, tall and handsome in jeans, boots, stripy cardigan, white V-neck shirt and grey jacket, carries a two-year-old girl in his arms, while he examines a stall of mushrooms displayed by a local forager. Michael Eavis, Glastonbury legend and local dairy farmer, stands talking to Julian, sporting a fleece jacket, very small, tight running shorts, bare, shaved legs and big, pristine running shoes.

Julian Temperley inspires extreme and divergent opinions, and some time after this day at Burrow Hill he and I had a major falling out that almost ended in legal action. But today, everyone here is friends. At 1 p.m., Julian calls us all over to a PA and microphone set up in front of his brandy distillery, and welcomes us to Apple Day at Burrow Hill.

Apple Day was started in 1990 by Sue King and Angela Clifford. Sue and Angela set up the charity Common Ground in 1983 with the late Roger Deakin, with the aim of exploring the relationship between nature and culture through art, poetry, film, photography and architecture. They also support 'local distinctiveness', a phrase they coined in the 1980s to help fight back against corporate commodification and standardization. Apple Day began as a single event in Covent Garden that sought to remind people of the huge diversity of traditional apple varieties, and to emphasize that even with a commercial market dominated by a few superstar varieties, there was still value in old orchards, especially if you consider value in a cultural and environmental sense rather than merely a commercial one.

By 2000 there were over 600 Apple Day events in the UK. As the first one was on 21 October, the tradition is to hold an event on the closest weekend to that date. In 2007 Sue and Angela published *The Apple Source Book*, a celebration of the apple's diversity containing pages of ideas for Apple Day events, and the day has now become a fixed part of the British calendar. The basic idea is simple, but can be interpreted in any number of ways. Apple Day events are now organized by the Women's Institute, National Trust properties, Wildlife Trusts, museums and galleries, horticultural societies, orchards, cider-makers, shops, restaurants, farmers' markets, schools and colleges. Apple Day is a balance between education and celebration, and that balance shifts depending on who is organizing the event, from displays of apple varieties through apple teas and apple-bobbing, to parties in cider farms.

'Sue Clifford and Angela King have done something wonderful,' says Julian from the stage, 'and I'm very proud that of all the Apple Days happening in the UK – and they've helped organize hundreds of them – they've chosen to come here today.'

Julian goes on to talk about his true passion – the Somerset cider brandy he makes here on the farm, and the bureaucratic battle he's had to fight to win permission to call it 'Somerset cider brandy', even though that's exactly what it is. Then it's time for him to introduce the guest of honour.

'Hugh Fearnley-Whittingstall was meant to be here, but he's gone fishing with the President of the Maldives. If this is part of the campaign against long-line tuna fishing, then we wish him all the best, because that is an abomination. But in his place, we have a very successful cheese farmer. He's been a judge on the Radio Four Food and Farming Awards. He's made cheeses including Little Wallop and Blue Monday. He had a previous life as bass guitarist in the 1990s megaband Blur. Please welcome Alex James.'

There are screams as Alex takes the mike, and laughter as one of the models theatrically pretends to faint near the front. Alex James is very posh. Twenty years ago, Alice Temperley was working in a cocktail bar in Covent Garden. Alex James was living above it and they got to know each other; she offered him some of her father's cider brandy. 'The twenty-year-old rivals anything from France, in fact anything I've ever sipped,' Alex now says. 'It's a reflection on how far we've come in twenty years. Things used to be so bad in Britain that people who eat snails and frogs' legs laughed at our food. Now, Britain is the most exciting place on the planet for good food and drink. For example, I never need to buy champagne again, now I can offer Julian's bottle-conditioned cider to my dinner guests. We're in the midst of a foodie revolution.'

As we walk away from the stage, Liz turns to me and says, 'Alex James just smiled at me.' There's an uncomfortable silence, before she adds, 'I still like you better.'

After this, we all head over to the orchards that flow from

the farm down the gentle slopes into the valley below. Now, the first of the year's crop has to be returned to the tree, in order to appease the spirits, in a tradition that may go back centuries, or may have just been made up. The marquee holding the Slow Food lunch stands at the top of the orchard's slopes, and that's as far as most people get, arrested by the soup, cheese, salad and meat arrayed on big central tables.

The minority of us who are curious or simply polite follow Julian Temperley, Alex James, the local vicar and a bunch of excited children down to one of the first trees, which is still laden with fruit. Julian instructs Alex to get down on his knees and hands him a bottle of the brand-new twenty-year-old cider brandy. Alex James uncorks the bottle and starts to pour it around the roots of the tree. 'Come on, all around the tree. The whole bottle,' snaps Julian, and a member of one of the most influential and revered bands of the last twenty years shuffles around in a circle on his knees, pouring away a bottle of very expensive brandy, while a vicar stands over him, solemnly intoning a prayer of thanks.

When the ceremony is over, we head back to the marquee and the scattered remains of a once-mighty buffet. I manage to find a few scraps of Montgomery Cheddar. I've been waiting for several weeks to try this with Julian's Pomona, and I'm not disappointed. The Pomona melts into the cheese, and they blend into one. After I've helped myself to the remaining crumbs of cheese and a few gherkins – you can always rely on gherkins to be left over – Bill points out the Common Ground ladies, Sue and Angela, so we pop over for a chat.

'The first Apple Day was in 1990, and we had the names of all the apple varieties in a border around the poster,' says Sue, or possibly Angela – the cider is starting to make its presence felt. 'We've stuck with it since then, because that's why we started it. We were aware of a culture that was disappearing. Lots of orchards were being grubbed up. The 1970s saw the rise of Golden Delicious and the British industry decided they wanted a monoculture as well to compete. So Apple Day was about preserving orchards. We think orchards are so inspirational because they show the marriage between people and nature. They do many different things and have a huge variety of species, most of them *down there*.' She stamps hard on the ground. 'Because the soil remains undisturbed for decades, there's so much life here. And the land has multiple uses. It's used for grazing, for flowers . . . Modern bush orchards are mono-cultural so not as good, but this way, the success of Burrow Hill shows it can be done.'

I mention the Marcher Apple Network stand in Much Marcle. 'Well, of course we gave them that idea, the apple road show. There were queues around the block.' One of them says all this, while the other remains silent.

Their ideas have taken root. In 2006, the Herefordshire Orchard Topic Group, Natural England and the Bulmer Foundation began a four-year study of six orchards in Herefordshire. They deliberately went for a cross section, from old traditional orchards to intensively cultivated bush orchards. The report: *Economic, Biodiversity, Resource Protection and Social Values of Orchards: A Study of Six Orchards by the Herefordshire Orchards Community Evaluation Project* was published in 2012. It was the first attempt to investigate the multiple values of orchards in the UK, and looked at economic, environmental and social sources of value, attempting to calculate a monetary sum for each.

Economically, the report obviously looked at the orchards' commercial profitability, but also at the benefit to the local economy from expenditure by orchard owners, and the spending of tourists who came to admire them.

Environmentally, they explored biodiversity and 'resource protection', the effects of carbon sequestration, soil quality and protection from pollution. They discovered a wide variety of species in the orchards and their hedgerows, just as Sue Clifford and Angela King had claimed. This was of course more prevalent in the traditional orchards, every one of which was home to species of fungi, lichen, mosses, worts and slime moulds (which are much prettier than they sound) that were either nationally endangered or uncommon in Herefordshire. The intensively cultivated bush orchards came into their own when their contribution to the protection of air, water and soil were examined. The more dense an orchard is, the more it becomes a carbon sink, taking carbon dioxide out of the atmosphere.

Finally, the social value of orchards was investigated by talking to people in the local communities and visiting tourists. The most important social values were the opportunities orchards provide to enjoy nature, and the simple appreciation of how attractive they are, followed by access for walking. Put simply, orchards make people happy. And while the economic value of happiness may be difficult or even impossible to calculate, it definitely exists.

The difficulties of putting a number on each of these values (not to mention the effects of how they interrelate) means that the overall monetary values calculated for each orchard were probably conservative. But even so, the calculations showed the overall monetary value of each orchard was at least double that of its commercial profit alone.

This report is a brilliant piece of work. We live in a boneheaded, short-sighted culture which has gone further than revering monetary value above any other form of value and is now making the case that monetary profit is the only source of value that matters at all. The Herefordshire orchards report not only shows that this is wrong, it translates its findings directly into pound signs for those who can't see anything else.

By the time we finish chatting, it's 2.30 and the crowd is starting to thin out. A man dressed as Farmer Palmer from the *Viz* cartoon walks slowly across the yard, placing each foot down carefully in front of the other, with great deliberation. Each step is an effort of will. Another man has just arrived, resplendent in a beret, goatee beard and rainbow-coloured sweater.

A pig emerges from a tiny Nissen hut and walks across the orchard, slow but determined, eating apples as it goes. With its deeply lined and furrowed face, it looks like a grumpier Patrick Moore or a kinder Bernard Ingham. Within thirty seconds it has a long train of child disciples, including Alex James's long-haired son, whom I swear he calls Artemis. Clouds of dust rise into the sunlight as the children pat its back. They stroke it. They embrace it. I'm confident that within minutes they'll be writing songs about it.

The adults show no such sentimentality: 'Yep, I reckon

it's about ready. Looks really tasty,' says one local. I grin, because this is the kind of joke city-dwelling wits such as myself would say to be cruel and upset their wives. But these guys don't look like they're joking.

The locals refer to Burrow Hill as 'posh cider'. But the majority of them have no qualms about turning up here on Apple Day, drinking as much as they can, as quickly as they can, ignoring the ceremonial aspect, stuffing their faces and then pissing off again. It's a lovely event, possibly my favourite of all the Apple Days I've been to.

After a few years exploring orchards, I've started to look forward to October the way other people do Christmas or the summer holidays. The orchards are still beautiful. It's exciting to feel the changing season rather than vaguely noticing that it's getting dark earlier outside the window. It's absurdly simple: whether we're eating free apple fritters next door to the original Bramley apple tree, buying cakes from the local Brownies, watching our dads make fools of themselves in an apple-pie-eating competition or a rock star shuffling on his knees around the roots of a tree, or simply tasting a new variety we've never seen before, the celebration of the apple harvest makes us happy. That has to count for something.

PART SIX

Transforming

October–November

'It's time to walk to the cider mill
Through air like apple wine,
And watch the moon rise over the hill,
stinging and hard and fine.'

Stephen Vincent Benét, *John Brown's Body*

19.

Life Expectancy

The sign on the old barn says 'Dunbar and Bunce, Thatchers'.

'There's a long relationship between thatchers and cider-making,' says Bill as he and I, Liz and Lisa get out of the car and cross the narrow lane to the barn. Here in the Somerset village of Kingsbury Episcopi – population 1,300 – there seems to be a relationship between everyone and cider-making.

Each year, Kingsbury holds a May festival. Once a gentle May Day fete, it has now grown into a weekend-long festival featuring live bands and a great deal of drinking. Last year, Tom Dunbar, thatcher and cider-maker, supplied a barrel of his stuff that disappeared in ten minutes, and quite a lot more cider had to be bought in from outside. Afterwards, someone pointed out that everyone had at least one apple tree growing somewhere on their land, so they decided to pool them together – cider apples, eaters, cookers, countless varieties. No one really knows what they have, or how it's going to turn out, but half the village has turned up to Tom's cider barn with their crop on this crisp October night. When Bill asked me if I wanted to come along and help make some cider, I started counting down the hours.

The barn consists of bare wooden pillars and beams overlaid with corrugated iron, the front completely open. A string of naked light bulbs sags from the rafters. This, plus the promise of various boxes of cider already made, welcomes us in.

⁂

We used to have an apple tree in the back garden of the house I grew up in between the ages of nine and eighteen. It produced big, fat cookers, and my mum warned us not to eat them because they were sour and would give us stomachache. My dad would half-heartedly harvest them, more to stop them falling and rotting on his vegetable patch than anything else, and my mum would bake an apple pie or two before losing interest. We just weren't big apple fans. Also, our household was one where any herbs, spices or seasonings beyond salt and pepper were considered pretentious, so my mum's apple pie, I realize now, would have been much improved by a dash of cinnamon.

We'd try to store the apples in the small kitchen cupboard, and they'd keep till the New Year. Before we moved to that modern redbrick house that was built on what used to be an orchard and had one surviving apple tree in the garden, we lived in an old Victorian stone terrace just up the hill, in the house my dad grew up in. We moved back there when my grandmother started getting ill, when I was seven, and lived with her for two years before she died. It had cavernous stone cellars that ran the length of the house and were always dry and cool and smelled of pavements after summer rain. My grandmother would store a box of apples down there at harvest time, and although I was very young and had forgotten this for years, when I think back now I remember the storage of apples in the cellar or garage over the winter being widespread – it was something everyone did. Now we get them in the supermarket all year round, there's no need. But the fact that this was common practice at least as recently as the 1970s is somehow comforting, even if it is a reminder of what a big issue food scarcity has been for most of our history.

Plants are continually respiring. Even after picking, apples continue to 'breathe', absorbing carbon dioxide and giving off oxygen plus a mix of ethyl and ester compounds that make their 'breath' smell, albeit in a way we find very appetizing. When apples are packed and stored, the concentration of their aromas evokes sweet perfume – mango, lemon and papaya, with an ethyl hint of nail polish. This respiration is another example of something we've long known but only recently understood. Scientists from the University of Leuven in Belgium and the attractively named European Synchrotron Radiation Facility only discovered the microscopic pathways through which apples breathe in 2008, using 3-D imaging and computer modelling. Even now, scientists have no idea how these pathways develop.

The discovery was an important one because as apples breathe, they age. After four or five months, even a carefully kept apple will soften and rot. Oxygen is an ageing agent, and the apple is continually producing oxygen. Now these respiration pathways can be measured, the optimal level of oxygen in apple storage facilities can be determined.

Apples lasted longer in my grandmother's cellar than they did in the little cupboard in the kitchen of our later, more modern house. In colder temperatures, the respiration slows down. Again, this is long-held folk wisdom – everyone knew it was the case that apples lasted longer in cooler temperatures, but didn't know why. Farmers used to bury apples in ice and they'd emerge crisp months later. In the diary of her life in India during the 1820s, Fanny Parkes wrote breathlessly of a shipment of apples that had arrived in Calcutta after a five-month sea voyage from the United States. Although ice was in such high demand (for cooling the drinks of the perpetually drunken British in India) that it sold for more money than the apples themselves, Parkes marvelled at the way it kept the fruit so fresh.

Shortly after the Second World War, researchers at East Malling began experimenting with new techniques of cold temperature storage, and then moved on to controlled atmosphere storage: if oxygen was the problem, what happens if you remove oxygen from the air and replace it with nitrogen? The answer is that the apple falls into a deep sleep. The production of ethylene gas, which promotes ripening, stops altogether and the apple remains in slumber, looking almost indistinguishable from the day it was picked – just like Snow White in her glass coffin.

I witnessed this at first hand on my visit to the Yakima Valley in Washington State. Given that the area provides 65 per cent of America's apple crop, the harvest in one corner of this vast country must be preserved for shipping across the continent at least until the harvest from the southern hemisphere is available six months later, ideally for an entire year if domestic growers want to compete with those imports. The apples that I saw being so carefully chosen and packed at the facility that Craig Campbell took us to in Yakima were going straight out on the road. The rest of the harvest was packed in wooden bins, 25 bushels to a bin. Each controlled atmosphere bunker is loaded with up to 2,500 bins, and there were 45 such bunkers in this one facility. We watched them being loaded by forklift, all the way up to the high ceiling. There are so many apples, and the smell of their perfume is so concentrated at this point, it makes you woozy. When a bunker is full, the doors are sealed and the lights go out. Oxygen is sucked out until it makes up just 4 per cent of the atmosphere and is replaced by nitrogen – and the apples sleep until the door opens again.

The fact that your 'fresh' apples may be at least a year old by the time you buy them in the supermarket horrifies many of us when we find out, and controlled atmosphere storage has become a favourite topic for consumer affairs TV programmes, magazines and websites. We experience a natural revulsion, a feeling that our fruit has been tampered with, as if it's undead, zombie fruit. There is, of course, nothing wrong with the fruit – it's been in suspended animation just like the bag of frozen peas in your freezer, and you're not freaked out by them, although some insist the fruit is both unripe and past its best after controlled atmosphere storage. But mostly it's the sudden discovery of this new technique, the fact that no one told us we were eating year-old fruit, that we find so unnerving.

We are, of course, complicit. Once, we accepted that apples were seasonal and their character changed depending on when they were ripe for harvest. Early season apples, picked in July or August, are fragrant and shiny, thin-skinned, and not especially rich in flavour, but very sweet and juicy. Most apples picked at this time bruise easily and don't keep well, meaning they're not attractive commercially. As the season progresses, autumn apples tend to be less bright, harder skinned and less fragrant, but the flavour is richer and they keep well if stored correctly. Cox's Orange Pippin and Blenheim Orange are typical examples. Finally, at the end of the year come apples with hard, dark, russeted skin, with low sugar but real depth and complexity of flavour, evoking honey and spice.

Now we expect apples always to look shiny and to crunch, crisp and fresh when we bite into them, and to give us a sweet, sugary rush of juice. We want early season apples, and we want them all year round. Gala, Braeburn and Golden Delicious are not true early season apples, but they share very similar flavour characteristics and are hardier and more reliable, which is why they're displacing more traditional, complex varieties.

Modern marketing demands consistency. When we buy a brand – be it HP Sauce, Pepsi or Budweiser – we expect each purchase to taste exactly the same as the last. That's the fundamental guarantee of what a brand is. And now we're extending that expectation beyond factory-made products to food and drink, where differences between regions, or from one year to the next, used to be celebrated. Wine has pulled off the trick of convincing people that it's acceptable that the character of different vintages can and should change. But even though hops, barley, apples, plums and cherries are subject to the same vicissitudes as grapes, we demand consistency from them.

And we take that constant supply and freshness for granted, until someone tells us a truth we don't want to hear. The truth is a simple one that we know but choose not to acknowledge: in autumn we're presented with a bounty of fruit, more than we could possibly eat. But winter is coming, its bleakness the start of another year before we'll have such bounty again, so we have to develop ways somehow to preserve the goodness of the fruit. In the thousands of years before controlled atmosphere storage, we developed many techniques for preserving and even enhancing the flavour and nutritional value of the apple far longer than the period the fresh fruit could be stored.

As well as baking apples into pies, they could be made into jams, jellies and chutneys, apple sauces and compotes, all of which could be better preserved than the fresh fruit. They could be dried or sugared. When I visited Uncle John's Cider Mill in Michigan, I learned that the bulk of the crop from one of America's biggest apple-growing states mostly goes to Gerber to make baby food.

Apart from the transformation of the apple itself into other apple-based products, we add it to all sorts of sweet and savoury dishes. The defining image of the medieval banquet – the pig's head with an apple in its mouth – reflects how strangely sublime these two ingredients are when brought together in any way, whether that's roast pork and apple sauce in the centre of an English Sunday lunch, chorizo cooked in sharp Asturian apple cider, the sliced apple and carved ham of a ploughman's lunch or the apple-wood-smoked sweetcure bacon of an American breakfast.

In *The Flavour Thesaurus* (2010), her culinary masterwork

that gives the best flavour pairings for over 200 different food ingredients, Niki Segnit notes that the apple can conjure up flavours of rose, damson, pear, pineapple, strawberry, rhubarb, plus spicy notes such as nutmeg and anise, dairy notes like butter, cream, cheese, plus hints of honey, wine, bubblegum, with the cyanide in the apple's pips contributing an almond note near the core. She recommends the apple as a flavour pairing with almond, anise, bacon, beetroot, black pudding, blackberry, blueberry, butternut squash, cabbage, carrot, celery, cinnamon, clove, coriander, hard cheese, hazelnut, horseradish, liver, mango, nutmeg, orange, peanut, pear, pineapple, pork, rose, sage, shellfish, soft cheese, vanilla, walnut and washed rind cheese. There are more ways to cook with an apple that might not last the summer than any chef will ever be able to fully explore.

But the oldest and most reliable way of conserving the goodness of the apple beyond the lifespan of the fresh fruit is to press it for its juice and turn it into cider. From here it can be turned into an array of different products. Among these, cider vinegar, which can be made deliberately or as the result of a cider fermentation going wrong, is currently being hailed as an elixir. Not only does it keep far longer than the apples from which it is made, historically it's been a vital aid to preserving other foodstuffs through the winter by pickling. On its own, it's the perfect ingredient to add to a salad dressing or marinade. If you can bring yourself to drink it neat, it's credited with having multiple health benefits. Experiments to verify these benefits are still inconclusive, but if you take a swig of cider vinegar straight with no chaser, the shock as it hits your palate and throat certainly makes you feel more alive.

The apple's versatility and adaptability, like many of its other attributes, serve to make it more important to us at the same time as making it seem ordinary through overfamiliarity. There's no better example of this than the apple's transformation into a drink ignored by most, derided by many, worshipped by a few and misunderstood by almost all.

20.

Scratting and Pressing

Work is already under way by the time Bill, Lisa, Liz and I arrive at Tom Dunbar's barn. People are anxiously rushing around with buckets, swapping full ones for empty. For many of the villagers, it's the first time they've made cider, and a sense of giddiness fills the air.

Tom's cider-making kit consists of two basic components: the mill and the press. The mill chops the fruit into a mush, but instead of mush we call it 'pomace', which sounds both more poetic and more scientific.

The importance of the mill presents a conundrum for cider historians. If you're making wine or grape juice, all you need is a press. But apples are hard and need to be broken up first. It's a simple enough task in theory, but get a bag of apples and try to make some juice, and you'll soon realize it's a bit more difficult in practice.

There have been various types of mill over the years. The most romantic is the big stone mill, a heavy wheel that sits in a circular trough and is pulled around by humans or horses to crush the apples to pomace, which is then shovelled out. This design dates back to the olive presses of Roman villas that were also used to press apples. The remains of stone cider mills are common around Herefordshire, Somerset and South Wales, most of them now serving as garden ornaments. Hellens Manor in Much Marcle still has a working one that presses pears to make perry every harvest time weekend.

Because cider is, in theory, such a straightforward drink to make, it's easy to assume that we've been enjoying it in Britain for thousands of years as a simple peasants' drink. But a stone mill remains a sizeable investment even today, and would have been beyond the reach of all but the richest in society for most of our history. A farm or estate such as Hellens might have had one, but if anyone else wanted to make cider on their own, they had a harder task. As apples age, they soften, and there are still traditional cidermakers around today who think damaged or softened fruit makes a better, more complex drink. Another alternative would be a mortar and pestle principle, breaking up apples in a barrel with a big lump of wood or rock. There's archaeological evidence of this method being used in Wales, and it was still in use on Welsh farms in the early twentieth century. But it would have been a time-consuming, labour-intensive process.

In the nineteenth century, mechanical mills replaced stone, and Tom's mill looks like an early example of this. It consists of a big wooden hopper, an inverted pyramid with a wide open top facing the ceiling, being fed a constant diet of apples that funnel down onto a wicked, spinning network of blades that in cider-making circles is known as the scratter. This is connected via a thick belt to an antique diesel engine that powers it, buzzing like an air raid and spouting gouts of blue smoke into the air.

From the scratter, the apple pulp, already brown thanks to its sudden, violent exposure to the air, is carried to the press. Again, there are many different designs of press. The traditional Somerset method is to build the pulp in layers of straw assembled inside a wooden frame, like we saw on the travelling press at Burrow Hill. A heavy wooden beam with giant iron screws then pushes down, squeezing out the juice and leaving a dry pomace which, on a traditional farm, is fed to the pigs living in the orchard. Later, the straw was replaced by rough cloths that bound the pulp together into fat layers known as cheeses, which hold the solids tight while allowing the juice to escape. In modern commercial cider-making the basic principle is unchanged, but hydraulics have replaced manually powered screws.

Tom's press, like his scratter, is a curious design. Instead of a solid frame, individual slats are placed together to form a square wooden frame. We spread the pomace in the middle and the frame gradually rises from the ground as we add more slats. As it gets higher, more slats are added, slotting together with tongue and groove to build a Jenga tower full of apple pulp.

The villagers are split into three teams: one loading the apples into the mill, one rushing buckets of brown mulch from the bottom of the scratter to the press, and one loading the pomace and building the press slat by slat.

I ask Tom and his wife, Amanda, about the curious design and antiquarian appearance of the kit.

Tom is in his late thirties, well built and strong without being fat or muscle-bound. He has short-cropped blond hair and a deep tan from spending the summer on Somerset roofs. And he has a beard. Tom's beard is not just any beard. It's thick, full-bodied and perfectly trimmed – not an embarrassing beardie beard, not a hipster beard, just a real man's beard. It stands out thick and proud where it should and doesn't bunch under his chin making him look like a bullfrog, like mine did the time I let it grow as long as his. It gives him the air of a polar explorer or a veteran of the Crimean War. It's a beard that refutes any suggestion of geekiness, inadequacy or weak-chinnedness that might be levelled at any of us lesser mortals who might dare to try to grow beards. As I gaze at Tom's magnificent beard, I realize I have no idea what he's been saying to me.

' . . . so we were looking for a mill to go with the old Hungarian wine press, and we were in Normandy, visiting friends. They have this website that's kind of like eBay for farmers, and we found the mill and that was perfect, but the bloke wanted to sell it together with the press. He was in his sixties, he didn't want to make cider any more, and neither did his kids or grandkids. It was filling up his barn and he'd bought a new caravan he wanted to get in there for the winter. We had this translator with us and through him the bloke said that if the press wasn't sold by the end of the next week, he'd chainsaw it for firewood. We couldn't bear that, so we bought both. Best two hundred quid we ever spent.'

Given that apple juice has to be pressed fresh before it can be turned into cider, it might seem as if I'm jumping the gun by going straight into a discussion of cider without having talked about fresh juice first. But there are two simple reasons for this. The first is that I'm biased towards cider. The second is that throughout history, most people shared my preference. How do I know? Because they didn't have any choice.

In nature, sugar is a scarce resource. It's a source of energy in a highly competitive world, and all manner of living creatures seek it out, from elephants to single-cell organisms. The family of yeasts known as *Saccharomyces* are microscopic and omnipresent. If you're reading this in a warm, summery climate, there are billions of them around you right now, and they're after one thing.

When *Saccharomyces* finds a source of sugar, and has a bit of oxygen to set it off, it goes on an orgy of gluttony and procreation. It reproduces at a rapid rate, consumes sugar and converts it into alcohol and carbon dioxide. When you leave a carton of fruit juice too long and it goes fizzy, and the carton stretches into an unnatural shape, that's the carbon dioxide by-product of fermentation, and your fruit juice is now alcoholic.* When you leave chocolate out for a few days and it gets a white, dusty coating, that's yeast, eating the sugar and reproducing. It's an entirely natural reaction, and one that's very difficult to prevent.†

Yeast is not only present on the skins of fruit – which is why grapes will start fermenting into wine as soon as they're crushed – in the apple it's already in the flesh itself. It lands on the blossom when it's being pollinated and sits dormant inside the growing fruit, until it softens or is

* Most freshly squeezed orange juice is naturally around 0.5% ABV, too low to cause any symptoms of drunkenness.

† Which is why, to paraphrase the late Bill Hicks, if you're a devout Christian and you're anti-alcohol, aren't you suggesting that God made a mistake?

crushed, giving the yeast the burst of oxygen it needs to begin its orgy. So throughout the entire history of apples being pressed for juice, whether you wanted it or not, you'd probably have ended up with cider rather than apple juice after a few days.

The process could be slowed down if you had somewhere cold to keep your juice – yeasts are sluggish at cold temperatures – and you could of course drink the juice fresh before the yeast has had a chance to get to work. Today, you can buy fresh, unpasteurized apple juice and if it's kept refrigerated it will last about a week before it starts fermenting, or 'going off'.

The revolution in preserving fresh juice came in the nineteenth century with pasteurization, a crude but effective tool. Exposing the juice to heat will kill off all bacteria within it, including yeast. Pasteurization usually happens after the juice has been packaged in sterile conditions, so it can extend the shelf life of juice from a few weeks to at least a year. But pasteurization is essentially cooking, and it changes the flavour of the juice, leaving us with a sliding scale of trade-offs between pure, fresh flavour and longevity and stability.

I never much liked the taste of apple juice when I was growing up, and now I understand why. Pasteurization on its own can be done gently, in a way that minimizes the impact on flavour and still leaves you with a decent juice. But combined with concentration, it can yield something that bears little resemblance to apple juice.

From a business point of view, not to mention an environmental one, reducing apple juice to a gooey concentrate makes an awful lot of sense. Liquid is heavy and expensive to ship. So we reduce the water content in fruit juice, ship it at far less cost to both the producer and the environment, and simply add the water back in at the other end.

The problem for the drinker is that juice from concentrate often tastes burnt and oxidized. Because it's stable at ambient temperatures, it then gets treated less carefully than fresh juice. And because big businesses are always looking to cut corners, you might find that the amount of water added to reconstitute the juice at the end is more than the amount taken out at the beginning. That's why, until recently, I believed that fresh apple juice tasted like cardboardy water. Happily, I've since discovered the good stuff.

In the United States, there's a very specific distinction between fresh apple juice and processed juice from concentrate. Cider as we know it in the UK was for a long time the most popular drink in America because it was easy for settlers to make. But when prohibition came into effect in 1920 and cider orchards were chopped down and replaced by other apple varieties, 'sweet cider' was redefined as fresh juice from recently pressed apples, often bought direct from the farm. When alcoholic cider made a comeback, it became known as 'hard cider' to differentiate it from sweet cider. Imagine wine being referred to as 'hard grape juice', and you can see why America's newly resurgent cider industry wants to reclaim their word. But the distinction as it stands between 'sweet cider' and 'apple juice' is an interesting way to indicate quality – the former is fresh-squeezed juice packaged on the farm, the latter is an industrial, processed product.

A great eating or fresh juice apple gives you a balance of

sweetness and acidity with little or no tannin. Tannin doesn't make an apple inedible, but it does make it challenging to many palates, whereas in cider it adds structure and body and stops the drink from being one-dimensional, which is why apples that are too astringent to eat often make great ciders. Some apples with just the right balance of sweetness, acidity and tannin can be used in multiple ways, and in theory at least there's nothing to stop you eating, cooking with or making cider with any variety of apple. It's simply a matter of taste – just ask Thoreau.

The cider industry uses a four-quadrant graph to evaluate apple characteristics, with the two axes being high to low acidity and high to low tannin.

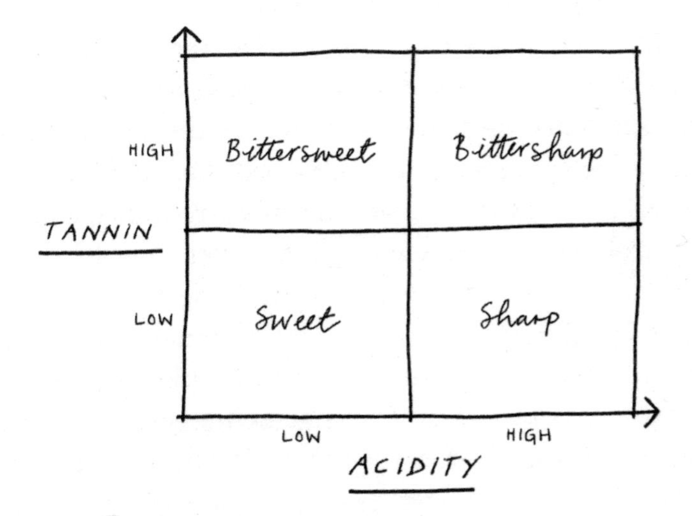

If we were to plot them on here, dessert apples would be bottom left – low in tannin and low in acidity, although some would drift towards the middle if they had a good balance of sweetness and acidity. Thoreau's sour crab apples and wild apples would be bottom right. The most prized cider apples are bittersweets, but they'll often be blended with sharps or bittersharps to achieve a pleasing balance of flavour.

In Herefordshire and Somerset, cider is made from named varieties of cider apple only, and they'll all be from the top two quadrants. But cider also has a long heritage in Kent, Sussex and Suffolk. Historically, it was much more common for eating apples to be grown here, closer to the big markets than the remote West, so dessert apples and maybe even culinary apples would be used in cider too. For the hardcore Somerset (or Devon or Cornwall) drinker, this stuff is not 'proper cider'. But Aspall, the oldest cider-maker in Britain, is based in Suffolk and has always used a blend of dessert apples with some bittersweets to add back-bone and structure. In County Armagh, Northern Ireland, long-established orchards that have traditionally focused on growing the Armagh Bramley's Seedling have reacted to a fall in demand by using the Bramley to make cider, and across the board the results are surprisingly good.

It's rare to find a single variety of apple that can offer the perfect balance on its own, but there are some. The Kingston Black, once on the verge of extinction, is now arguably the most revered cider apple in the world, a perfect all-rounder. My own favourite is Dabinett, supposedly discovered as a wilding or 'gribble' in a hedge at Middle Lambrook, South Somerset, in the early 1900s by a Mr William Dabinett, and quickly propagated across the county. The fruit is small, seductively red, bordering on pink, with stripes of pale green. The cider it creates is medium in its sweetness and has a soft astringency that often conjures up hints of pepper and spice. It was the first single apple variety I was able to identify in a blind cider-tasting, and it is for me the equivalent of Syrah or Gewürztraminer in wine (great cider can combine the flavour characteristics of both red and white wine).

My exploration of the possibilities of single varieties brought about a reconciliation between me and fresh apple juice. After accepting that I could no longer eat apples, but that drinking cider was fine, I wasn't sure where unfermented juice sat on the scale, and my earlier experiences with watery cardboard meant I wasn't in any rush to find out.

And then, while in Putley, I visited Glebe Farm, about a mile down the road from Dragon Orchard in the parish of Aylton. Here I discovered that fresh, unpasteurized apple juice only provokes the faintest tingle in my throat, and that fresh juice made from named apple varieties has its own delights.

The big difference from cider is the lack of tannin, allowing a simple balance, a dance between sweetness and acidity that reminds me at times of a soft, fruity white wine. Lord Lambourne is thin and clean with hints of apple and grapefruit, while Worcester Pearmain is broad and sweet, a big, friendly juice that gives your palate a warm, clumsy hug. But my favourite is Discovery. The juice is paler than the others, the colour of fresh lemonade. It's drier too, with a faint hint of tannin and some nice, delicate high notes.

In Tom Dunbar's barn, the press rises, albeit not quite in the right direction. A man who looks like a schoolteacher opens his mouth, and sounds like anything but when he says, 'You musta bin pissed when you built thaaaat!'

''E' leaning fair! 'E' leaning fair to the left!' cackles another man.

And so 'e be.

It feels good to be making cider in Somerset – a balancing of the books. Ever since I first started writing about it, people have always asked me which cider-making region I prefer, and I'm afraid it's Herefordshire. Read the history books, and that's where all the great advances in cider-making – not to mention apple-growing – were made. Herefordshire was where the activities of the Woolhope Club kick-started the salvation of the British apple industry. Today, when people ask me for a list of my favourite ciders, I get to about five and realize they're all from Herefordshire and I have to make a deliberate effort to include some from the West Country. When I admit my preference, this is inevitably seen as some kind of criticism of Somerset, and its supporters insist on telling me its many attributes and wonderful ciders. And I agree. I know this. Preferring Herefordshire cider doesn't mean I'm unaware or unappreciative of Somerset – I am allowed to like both.* It's common for cider-lovers to describe Somerset as the

* I encountered the same binary thinking from beer-lovers when I first started writing about cider. 'Oh, don't you like beer any more then?' And both beer- and cider-lovers seem to think that if they ever catch me drinking a glass of wine the betrayal is severe enough to be blackmail material. These are all more than drinks. For their most passionate adherents, they are lifestyle choices and badges of identity.

heart and soul of cider, and I agree. But if that's the case, I'd suggest Herefordshire is the brain, the intellect of cider. You need both brain and heart to prosper, and Somerset and Herefordshire are complementary.

But I have spent a lot of time in this apple year in Herefordshire, and not enough in Somerset. There's something magical in the air whenever I come down here on cider business. Normal rules don't seem to apply. I may not believe Somerset *makes* the best cider, but I do believe it's the best place to *drink* cider. And now I have a chance to explore this by doing both simultaneously.

Tom's cider press looks more than a little tipsy. Squeezed by the weight of the pomace above, some juice is already trickling out from the bottom of the cheese into a waiting bucket, somehow adding to the impression of dissolution. Nevertheless, the final slats eventually go on and the last bucket of pomace from the scratter is spread over the top. A young, blonde, Barbour-clad woman called Sam is persuaded to take off her shoes and socks and climb gingerly onto the top to trample down the apple mush to be level with the top of the press. Many of the men present seem curiously reluctant to let her stop and climb back down.

When they finally admit that she's done her job and they have no excuse, Sam departs the press. We place wooden boards across the top, sealing in the pomace, and the pressing begins. Again, it's an unusual but clever system: a long metal arm attaches to the screw and curves out and down from the top of the press to around shoulder height. From here, it can be swept in a broad arc from left to right. Somehow, via a system of metal weights that rise and fall in a singsong clatter and clacker, this back and forth motion is translated into a continuous turning of the big metal screw at the top of the press which slowly pushes down to crush the pomace.

It's gradual work that requires patience. The screw comes down slowly, but soon the big plastic bucket at the bottom of the press is being replaced every fifteen minutes, the juice a deep, rich, ruby red. I take my turn on the press, and to begin with it's easier than I thought. As the screw lowers, the top slats are taken away again, one by one, but the structure is definitely starting to look even more lop-sided. 'It's on the piss, definitely,' says Tom. 'Just like us!'

The effect of having a press that is itself pissed is that the screw is slightly off to one side, pushing down harder on the left than on the right. The resistance starts to rise as the pressure of the compacted apples grows beneath the board. Soon we're working it in teams of two, pushing and pulling towards each other.

After the nervous anticipation at the start, the work slows to a steady rhythm. Gravity and pressure cannot be rushed. Anyone not on lever or bucket duty stands around watching, lighting up the air with jokes and laughter. The barn is thick with the aroma of apples, making stomachs rumble, and Amanda starts to pile up a wide trestle table with long, chestnut-coloured loaves of bread, cheese and a variety of neatly labelled homemade sausages. We all dig in and help ourselves.

Apart from Bill and Lisa, I've never met any of these people before, and I can't believe how welcoming and relaxed the evening is turning out to be. The whole event is about neighbourly generosity, a community coming together, each bringing what they have, to make something they'll get a collective benefit from. They're doing it after dark because they all have day jobs, but the clear, starlit night gives it an air of celebration. And though it's perfectly legal for anyone to make less than 7,000 litres of cider, the night and the spartan barn also give this gathering the distinct air of an illicit still.

Each bucket of rich, brandy-coloured apple juice is poured via a simple funnel and sieve into new oak wine barrels and a couple of whisky barrels, one just bought from the Laphroaig distillery. The yeast from the fruit is still alive in the juice, ready to go now it's been released. It will ferment slowly in the barrels over the winter, and that's it – that's how you make cider. Once the barrels are full, it's just a matter of waiting, give or take the odd peek to make sure everything's OK.

This is how cider has been made for centuries – the principles are pretty much unchanged. For much of the last millennium, cider rivalled beer as Britain's national drink. It hasn't been celebrated anywhere near as much as beer, has had a fraction of the amount written about it that beer has, but has been consumed so heartily that – just like the apple – many of us think it is indigenous to Britain, and express surprise that it's made anywhere else.

But this is what cider does, just like the apples from which it's made. When it finds a home, it settles in, becomes insular and keeps to itself, slightly off the map, as if it's always been there and only there. When Bill and I travelled around Europe as part of our attempt to pull together the first ever global overview of cider styles, we found long-standing traditions in Asturias, Normandy and Frankfurt. In each place, we were asked why, as Englishmen, we were interested in cider. When we replied that Britain made cider too – that in fact cider is so popular in Britain our nation accounted for over half of global production until recently – reactions ranged from incredulity to flat disbelief. Cider is out of sight, in the countryside. Even today, no one knows how many cider-makers there are in Britain, because if you still make it the traditional way, as a sideline on a farm, you don't even have to declare it to HMRC – or even register with them – if you're making less than 7,000 litres a year.

This gives cider a delicious air of mystery for those who love it. I've found many of the best cider-makers in the country to be a truculent bunch, not too bothered about commerce, plain speaking and mulish. If you want to taste the best ciders Britain produces, chances are you'll have to go looking for them in the regions where they're made. The downside of this is that most us remain unaware of just how great cider can be.

The activity in Tom Dunbar's barn is a perfect example of the basic cider-making process, but there are variations within it, and no two cider-makers will ever agree on what's best. Yes, there are natural yeasts in the apple that will ferment the juice into cider if they are left to it. But wild yeast also adds its own flavour characteristics. Often these are what drinks writers describe as farmyard – spicy, funky, ripe earthy notes that can evoke associations like manure, leather, cheese or sweaty horses. You can make a cleaner cider by adding champagne yeast, which overpowers the natural yeast and leaves a much more delicate, biscuity or grapey contribution to flavour.

There are also different approaches to how you blend apples. Here at Tom Dunbar's, we don't know what the varieties are so we just stick them all in and hope for the best. A traditional Somerset cider-maker might judge it by sight, gathering a huge pile of different varieties as we saw in the farmyard at Burrow Hill and getting an instinctive blend of different colours as the apples go into the mill. Tom Oliver of Oliver's Cider and Perry in Herefordshire ferments each apple variety separately in wooden barrels, each of which has its own characteristics from the wood and the microflora within it. He then tastes each barrel and painstakingly blends them, in his words, 'bringing out what the cider wants to be'.

Some cider-makers believe that whatever you choose to do, once it's in the barrel fermenting, you stick with it whatever happens. Others take a careful approach and check and measure what's happening over the fermentation period, adding nutrients to keep the yeast healthy wherever necessary. Some add an extra dimension by finishing the cider in whisky barrels, like Tom Dunbar is doing with some of his. Others regard this as cheating.

Get two cider-makers in one room and you'll have three different sets of opinions about cider. One cider-maker told me that cider must be drunk within a year and that there is no way it can be kept longer, no matter what. Half an hour later, I was introduced to another cider-maker, who asked if I'd prefer to try his fresh cider or his aged vintage stuff first.

In Britain, we tend to think of cider in relation to beer,

and in America many drinkers even refer to it as 'cider beer'. But true cider has as much, if not more, in common with wine. Strength-wise, natural cider sits handily between the two: the average strength of beer in Britain is 4.2 per cent alcohol by volume (ABV); wine was 13 per cent ABV and is now 14 per cent. Natural, unadulterated cider usually ferments to somewhere between 7 per cent ABV and 10 per cent ABV.

Cider's similarity to wine becomes much clearer when you encounter cider in Normandy or Brittany. Cider-makers there talk passionately about *terroir*, and package their products in tall 750ml bottles. Drive around Normandy, and wherever you go you'll pass farms advertising cider for sale, but some of them express bemusement that people want to drink it. If you distil wine, you get brandy, the finest example of which is Cognac. In the Pays d'Auge, many of the people who make great cider are only doing so as a precursor to creating the region's true speciality: Calvados.

Most Normandy cider-makers produce superb cider, for the simple reason that only with the best cider can you make the best Calvados. Freshly distilled apple brandy is clear when it comes off the still, fiery and raw on the palate, but the good stuff still retains the fruity character of the best cider, which can in turn only be made from the best apples. For apple spirit to earn the Calvados name, it has to be aged in oak barrels for at least three years, where it becomes softer and the apple and spirit mingle with the wood. As it ages, it acquires spiciness and butteriness, and the apple character is a mercurial ghost, moving in and out of your perception.

The possibilities and permutations don't end there. Just as the brandy-makers of Charentes blend freshly distilled wine spirit with fresh grape juice to create the regional speciality Pineau, so you can blend raw apple *eau de vie* with fresh apple juice, and age that in wood for three years, to create Pommeau. At around 17 per cent ABV, Pommeau has more in common with sherry or port than anything else, forms a perfect *aperitif* or *digestif*, and is even better with cheese than the cider from which it's made.

Calvados and Pommeau are traditional products of the Pays d'Auge that are now spreading to other countries as different cider regions finally start talking to each other and sharing ideas. And they've now been joined by a new delicacy that carries on the Normandy tradition from across the Atlantic. *Cidre de glace*, or ice cider, was born in Canada when Christian Barthomeuf, a French émigré who originally tried his hand at ice wine, noticed some unidentified varieties of apple that stayed on the bough through the depths of the Quebecois winter. At temperatures of minus 15 degrees Celsius, the wind and cold 'cooked' the apples as well as freezing them. Water would eventually erupt out of them, leaving shrivelled fruit with concentrated juice. German Eiswein, a dessert wine created by leaving grapes to freeze on the vine throughout winter, already had a long history and was being made successfully in neighbouring Ontario, so Barthomeuf tried a similar idea with apples. In 1994 he collaborated with a neighbour, François Pouliot, to create Neige, the first commercial ice cider brand. The freezing process concentrates the acidity as well as the sweetness, so instead of a sticky, sickly dessert wine quality, the flavour is intense, vivid and energizing. When I first tasted it, I wrote, 'It's like drinking starlight.' Even the wine-makers at Château d'Yquem agree that *Neige Récolte d'Hiver* (winter harvest) – made only with those apples that stay on the tree through the winter, adding an extra layer of stewed fruit character – rivals the very best dessert wines.

First juice, then cider, then Calvados, Pommeau and ice cider show how a fruit with a six-month shelf life, a fruit facing severe price pressure in Europe because of a glut of supply, can be turned into a value chain that lasts indefinitely. Some North American orchardists have turned to cider-making because they can no longer sell their apples profitably as fruit for eating. Initially using only the dessert fruit they'd been growing exclusively since prohibition, they lacked the structure-giving properties of tannin and had to show great skill to prevent their ciders from being bland and monotone. The best – some fermented up to 10 per cent ABV – are far more reminiscent of a good Riesling than a bottle of Magners. These American growers are passionate about their apples, and believe that cider-making allows them to show the fruit at its absolute best, that cider is the finest expression of a great cultivar. This potential, newly explored, is echoed by the name of the company that makes the remarkable *Neige – La Face* Cachée de la Pomme (the hidden face of the apple).

Even though it's late in Tom Dunbar's barn, there are lots of children running around. If this were in London, they would have been brought in pushchairs, and by this time of night the air would be full of fractious screaming from kids kept up too late past their bedtime. Here they're left to run around the barn freely, and the excitement has affected them: they're burning off energy, chilling out. I don't think the time for me to abandon London for the sticks is anywhere near – not just yet – but when it comes, this moment will be one of the main arguments in favour.

Bill has spent the whole evening up on hay bales, tables, in corners, snapping away to capture the action and the atmosphere. Now he comes over and asks if I've noticed Tom's beard.

I reply that I have.

'He's just really cool, isn't he? He's the kind of bloke you'd want to work for.'

I realize that both Bill and I are hero-worshipping Tom. We're both a little bit in love with him. But Bill's not wrong – there are thirty people here on a Saturday night, in Tom's barn, quite happily doing whatever he tells them to do.

And now here comes the man himself. I attempt to change the subject, but Bill has other ideas.

'We were both just really admiring your beard!' says Bill as the great man joins us.

I cringe inwardly with embarrassment. But it soon becomes clear that we're not the first people to praise Tom for his facial hair prowess. He tells us that mere boys come up to him and say things like, 'Hey mister, do you think I might grow a beard like that one day?' In one pub frequented by squaddies, the drinkers were adamant that he was a member of the SAS, refusing all protestations to the contrary with a knowing 'Of course, you *would* say that.'

As Tom chats about beard-fame, the rest-gaps between

sessions working the press become longer and longer, the flow of juice slower, until finally, via some unspoken agreement, we give up.

Outside on the narrow, high-hedged lane (I can't find the toilets in the barn, despite repeatedly being given directions), as scraps of cloud scud across the dense stars, I try to work out why this open-fronted barn, with its lights pouring out into the night, stirs something so deep in my soul. And I realize that it's because we see re-creations of it every year, in our homes and outside churches. It's the framing of a nativity scene.

Would it be too much to argue that there's something divine being born inside this barn tonight? I suppose we'll have to wait until a few weeks after Easter to find out.

PART SEVEN
Slumbering

November–January

'Here's to thee, old apple tree,
Whence thou mayst bud
And whence thou mayst blow!
And whence thou mayst bear apples enow!
Hats full! Caps full!
Bushel – bushel – sacks full,
And my pockets full too! Huzza!'

Traditional Wassail song from
South Hams of Devon, 1871

21.

Winter Orchard

For anyone who takes joy from the display nature puts on throughout the year, the few brief weeks when pink turns to white are show-time in the orchard. Then, as the fruit grows, so does our sense of anticipation of harvest. The harvest itself celebrates ripeness and bounty, and ultimately the surrender of all our senses to the alluring fruit. The blossom and the fruit are the two products of the tree, the rewards for tending it. Ancient myths told of both appearing simultaneously on the enchanted trees of Paradise. The artists who illustrate the various volumes of *Pomona* depict fruit and blossom together because they can.

By contrast, the winter tree is naked. An orchard in summer is a concert of trees, each adding its part to the beauty of the whole. In winter, each dormant tree stands alone, apart from the others, lost in its own dream.

But the winter apple tree has its own proud beauty. The architecture of the tree is sketched bold against the landscape. The clawed branches reach up and out, cradling the sky, and the past year's new growth shoots straight up in a windswept mane. Close up, the wizened bark makes the tree venerable, the branches kinked by knots like arthritic fingers. Lichen furs the trunk and boughs, thick and mossy, and when the frost hits, a million tiny needles give the surface of the whole tree a silver halo. As Liberty Hyde Bailey reflected almost a century ago, the winter apple tree 'has none of the sleekness of many horticultural forms . . . [but] . . . it presents forms to attract the artist. Even when gnarly and broken, it does not convey an impression of decrepitude and decay but rather of a hardy old character bearing his burdens.'

In a few months, the tree will look young again. Its stark, crooked winter frame is as temporary as its callow spring blossom, a fleeting illusion created by our persistent delusion that time is linear rather than circular.

I always used to feel sad about leaf fall. I think everyone does. My music collection is dotted with odd songs themed after months of the year. 'Little Bird' by Goldfrapp, with its refrain of 'July-ly-ly', is woozy and luscious. 'August' by Adult Net is the sound of endless days on the beach. 'September' by David Sylvian is short and melancholy, while 'October' by U2 is sparse and bleak (and surprisingly good given that Bono sings it), a perfect partner to Cousteau's 'The Last Good Day of the Year', whose lyrics speak of fragrant summer light giving way to autumn's burnished fingers and the cold wind breathing down your neck. Autumn feels like the end of the party. For those of us who are out of touch with the wheel of the year, there's no big, defiant celebration like Christmas, no combination of reflection and looking forward like New Year, just stuff dying, the nights drawing in, the beach huts and concessions closing up, the barbecue cover going on and the garden furniture being stowed in the cellar.

But both leaf fall and the period of dormancy that follows are essential for deciduous trees. Again, we perceive leaf fall as a symbol of ageing and dying. But it's a practical – and deliberate – step from the tree's point of view. The broad, literal meaning of the word 'deciduous' is 'the dropping of a part that is no longer needed'. A child's milk teeth, and the antlers of some deer, are also deciduous. In hot climates, deciduous trees shed their leaves during the dry season to conserve water. In temperate climates, they do the same thing during the cold season. Branches heavy with leaves are more likely to snap off under the weight of a heavy snowfall than the slim needles of pine trees. Leaves also become damaged by insects and scab, and can provide sanctuary for pests. It can be less expensive in energy terms for a tree to shed its leaves and grow new ones rather than constantly trying to repair them.

So as winter approaches, the apple, like other deciduous trees, cuts off the supply of chlorophyll to its leaves. The vivid reds and golds we see in autumn are there throughout the whole leafy period, but as the tree takes the chlorophyll back into its roots to break down and reuse, these shades become visible. When there's nothing left in the leaves worth salvaging, the tree cuts them off via a process known as abscission. Like hibernating mammals, it sharply reduces its rate of respiration. It continues to take in nitrogen, phosphate, magnesium and other minerals to stay alive, but ceases to grow in any way.

The apple tree needs its dormant period, just like we do. Harvest is a busy and stressful time, and it needs to recover. 'We all need a time to do, and a time to rest,' Norman Stanier told me when we were admiring his freshly wakened trees back in May. 'The trees need to stop growing to store up their vigour for next year. If we don't rest, we get stressed, and so do they. They're the harbingers of the season. They need both hot and cold, light and dark, the duality. Just like us.'

'People don't realize that when you're looking for the right climate for apples, it's not the height of summer temperatures that matters, but how low it goes in winter,' explained Tim Biddlecombe when I visited him at Brogdale. He introduced me to the concept of 'chill units'. A single chill unit is one hour at a temperature between zero and seven degrees. A dormant apple tree needs a thousand chill units each winter to enter a proper state of dormancy. If it doesn't get a good winter's sleep, its performance is just like yours when you've been suffering from insomnia. The bud burst is erratic, with blossom appearing on some parts of the branch while it remains tightly sealed further up. The terrible apple harvest of 2013 was attributed to a winter that was too warm and too wet rather than a summer that was too cold.

'They don't like wet feet,' Norman told me. 'You can see them looking miserable. When it's warm and wet, they're wasting energy, and that makes for a bad harvest. And if they blossom too early, the chances of a late frost killing off the blossom entirely are greatly increased.'

As soon as I learn about chill units, I become obsessed by them. I'm finishing this book over a winter in which the weather has become entirely unpredictable, even if our relationship with it is scripted and routine. In November, the *Daily Express*, as it does every single year, terrifies its readers with warnings of twelve weeks of arctic temperatures that never occur. Floods wash away parts of the North, politicians put on wellingtons and carefully but inadequately practised expressions of concern, promise to do 'whatever it takes' to make things better, and then do nothing at all. By late January, the Boxing Day floods are still the lead story in the news, with more to come, and Brogdale has recorded only 445 chill units, compared with 750 by the same day the year before.

While the trees try to sleep, there's still plenty of work to be done in the orchard. The first task is to clean up the orchard floor. Fallen leaves and fruit are a perfect haven and breeding ground for fungal spores, and should either be removed or broken up and mulched into the soil. In theory, removing every fallen leaf means there'll be no scab the following year.

It's also time for at least some pruning. Root pruning involves cutting around the tree with a sharp blade to crop the roots and keep the size of the tree reined back, but it has to be done while the tree is dormant or it will limit the size of the fruit too.

Pruning of the branches also traditionally takes place in winter because it's a relatively quiet time. But pruning for delicate shaping to maximize light and air in the tree is not necessarily a good idea at this time of year. Prune when the tree is dormant and you get the most vigorous regrowth. This may be what you want, but if you're trying to stop the tree from growing in a certain way, cutting it back in winter ultimately has the opposite effect to the one you're trying to achieve. Also, by cutting a tree you leave an open wound, which can make it vulnerable to apple canker, a fungal disease that attacks the bark and eventually kills branches, or even whole trees. Favourite apple varieties such as Jazz and Gala are particularly susceptible to apple canker, and pruning in winter leaves the tree with many openings for canker's arrival in spring. For these reasons, at the National Fruit Collection in Brogdale, pruning is being deferred to later in the year.

Despite the risks, big, structural pruning is best done in winter, when you can see the shape of the tree clearly and are able to make big, deep cuts that will guide it over years rather than one growing season. Pruning is a very important task – one of three most labour-intensive jobs in the orchard. So having mastered the art of chip bud grafting, and picked a few apples from a tripod ladder, I now decide to complete the set and find someone who will let me hack away at their prize apple trees with a saw.

Tom Oliver is one of the three or four best cider-makers in the world – quite possibly the single best. He is also unique in that among the many different cantankerous factions among cider-makers, everyone respects him and likes him. His round face and serene smile give him the countenance of a happy Buddha. He spends most of his time on his family's sheep farm at Ocle Pychard, another one of those Herefordshire spots that sounds like a made-up place, but isn't. He juggles cider-making and the rescue and propagation of rare perry pear varieties with tasks such as lambing. When he can get away from the farm, he also has a career in the music industry as a widely respected sound engineer who, among other commitments, acts as tour manager around the world to the Proclaimers. At a recent cider convention in Portland, Oregon, this quietly spoken middle-aged man was accorded rock-star treatment himself, constantly mobbed by glamorous young men and women desperate to have their photographs taken with him.

I often characterize various talented producers in brewing and in cider-making either as craftsmen, curators, mad scientists or chefs. Tom for me is the rarest breed of all: an artist. I can't think of anyone whose trees I would rather mutilate.

Tom picks me up from Ledbury train station on a sharply clear Monday evening in January. He takes me to the Annual General Meeting of the Three Counties Cider and Perry Association, where he is again elected president. One item of business during the meeting concerns my old friends, the Marcher Apple Network, who I watched attempting to identify apples at Hellens Manor back in October. The average age of the members is advancing to the point where they're now struggling to carry out their duties, and Tom has been asked to make an appeal for young blood and new life. I feel a sudden rush of love for the ruminative elders who seemed carved from wood themselves, and allow a fantasy of living in the Welsh Marches, planting my own orchard of rare varieties, making a bit of cider and helping run organizations like the Marcher Apple Network to carry me through the rest of the agenda and all the way out of Any Other Business and into the simple supper of cheese, ham, bread and cider that follows.

The next morning I get the chance to play out a little more of that fantasy. After breakfast, we leave the farmhouse and step out into a silver world. The clear night has left a thick, hoary frost on the ground, and the dawn has brought an icy mist to keep it in place. Every blade of grass is outlined individually with frosty fur. A thick layer of it sits on top of the trimmed hedges and rimes the trees. After a few minutes swishing through the long grass, the tops of my boots are encased in solid ice. The very air is silver, sealing us in monochrome. This is one of the best aspects of getting in touch with the apple year: back at home, this would be just another grey day outside the window, with nothing interesting to look at. Instead, even one of the darkest days of a dull winter has been transformed into something wonderful.

Tom buys in a lot of his apples from other local farmers, but about twenty years ago he planted a small orchard of his own, focusing on rare and interesting varieties. In the early years he pruned them for future growth, but by his own admission he then 'just let them get on with it'. These are standard-size trees, widely spaced, and it's nice to see a more relaxed approach after the ever more intensive practices of commercial fruit growers. But now the trees definitely need some work. Each one is good and strong, but the ground is fairly flat, soft and boggy in places. Last year Tom lost a couple of trees that went over under their own weight. To the west stands a dense coppice of tall trees, which means Tom's trees all stretch south-east, towards the sun in the first half of the day. Many of them now have thick boughs pointing long and straight towards a point where the mid-morning sun is now trying unsuccessfully to burn through the mist.

'You leave it at the beginning, because those branches are giving you loads of fruit,' says Tom. 'But at some stage you have to accept it's pulling the tree over and inhibiting new growth.' Left to grow in the direction they want to, the trees' yearning towards the light will most likely kill some of them prematurely. It's fascinating – and a little daunting – that as well as making the tree behave the way we'd like it to, our intervention is also helping it to survive.

We must be delicate and well judged. Prune too aggressively, and in Tom's words, 'The tree will rebel.' If we want to avoid the tree producing far too much wood in response, we have to take out no more than one or two big branches a year. As well as branches that overbalance the tree, we're looking for anything dead or diseased, any branches that cross, and any that are crowding each other out. From a fruit-growing perspective, there's little point allowing a tree to expend its energy on growing a big limb that sits directly under another limb of the same size, taking all the sun so the one beneath doesn't get any light.

'When the tree first starts growing, you're looking for balance around the centre,' says Tom. 'Say you got four branches coming out at the same height, you'd get rid of two of them because otherwise they're putting too much pressure on the tree, taking too much out. The perfect shape would trace a path like a helter-skelter, the big branches coming out in a spiral as you go up the trunk.'

Tom is an instinctive teacher. After explaining the basic principles, he takes me to the nearest tree and asks me what I think we should cut. This first tree has one low, heavy limb dragging the ground, wool in its claws from where the sheep have rubbed against it. Another thick branch points to the south-east so long and straight it seems to be making a bid to escape the shackles of the tree of its own accord. I point to these two and Tom nods, almost imperceptibly. 'Where should we cut them?' he asks.

Every single person who comes to pruning for the first time is too timid, terrified of hurting the tree. Even after the principles have been explained, there's a deep-seated guilt about vandalizing the trees, a fear that they may not recover from being disfigured. But if a cut is necessary for the tree's welfare, it usually means it's best to go in hard. If a bough is pulling the tree out of shape, or if it's in shadow, will half a bough be any different? As we discuss possible cutting points, they invariably move back towards the trunk itself.

Of course, you have to be very careful. As a novice, you tend to focus on what you're cutting off, but the only thing that matters here is what you leave behind. Our main concern is to minimize any distress to the tree, and that means pruning often has to be a multi-stage process.

Tom talks me through a method known as the Dutch cut. First, we find a spot four to six inches along from where we want to cut the bough from the trunk. Here, from underneath the bough, we saw upwards until we're about a third of the way through, or until the weight of the bough closes the cut and threatens to trap the blade. Then, we move another two or three inches further out, and start sawing from the top. Now the weight of the bough is pulling the second cut open, stretching the fibres on the bottom half. As soon as the weight of the bough is stronger than the integrity of the wood around our cut, it starts to split and snap. Without our first underneath cut, at this stage the wood could split all the way back to and into the trunk, severely damaging it. Instead, the split reaches the underneath cut and the bough snaps off. Having removed most of the weight, we can now focus on making a third cut that's clean and simple and minimizes any damage to the tree. Back at the trunk – where we actually wanted to take off the bough – we only have to worry about a few inches of wood. Angling slightly diagonally down and out, with a new and very sharp saw, I make a clean, smooth cut, the angle meaning water will run off rather than gathering; the smoothness enabling a quick healing process and minimizing the chances of infection.

Once we pull the felled boughs away, the trees immediately look better, just like you do when you've had a sharp haircut. Tom will saw the thick wood into lengths and allow it to dry out for a year before using it for firewood.

After I've lopped off four or five big boughs from various trees, Tom takes me to a tree that has a huge, twisted limb that starts near the base of the trunk but stretches out like a pointing finger, and also sends a bold shoot vertically upwards into the canopy. I look at what we're proposing to cut away, and it seems like an awfully large proportion of the tree. Should we cut off the long pointy bit and allow the tree to put everything into the spirited shoot that's made its way twenty feet up into the canopy? Tom shrugs. These vertical branches, known as water shoots when they first appear, tend not to bear fruit and just suck out energy. But can we really take the whole lot off?

And then, instead of looking at the amount we're proposing to take away, I look at the tree and try to see how it would look without these branches. Immediately, I know I've got it. I think of Michelangelo, and the almost certainly untrue story of the Pope asking him how he created his sculpture of David, to which he supposedly replied, 'You just chip away everything that doesn't look like David.' Like all such parables, the story survives because it's useful. Sawing away any big branches that don't look like part of a healthy apple tree is the secret to successful pruning, but you have to keep your focus on the tree, not the parts you're cutting away. I wonder if Michelangelo ever used the shards of marble on his studio floor for anything as useful as heating?

Apart from tidying up the orchard and the trees, winter is also a busy time for tree nurseries. Dormancy is a good stage to undertake many forms of grafting, as well as for planting new trees.

In 1901, Frank P. Matthews opened his first nursery in Harlington, Middlesex. Natural fertilizer was easy to come by from barge-loads of manure taken from London's stables. Business grew, but was then stopped by two world wars, when priority had to be given to the cultivation of cereals and potatoes. Just as recovery from the Second World War was under way, the land on which the nursery stood was claimed for new developments around Heathrow Airport. The business moved to its current site on the banks of the River Teme in Tenbury Wells, Worcestershire, and gradually expanded across the small valley it occupied. It's now the largest fruit tree nursery in the UK, supplying orchards and garden centres with over half a million fruit and ornamental trees every year.

Today, the company is run by Nick Dunn, Frank's grandson and current chairman of the Royal Horticultural Society's Fruit, Vegetable and Herb Committee. Nick is in his fifties, relaxed and supremely happy to have taken on the family business. He tells me several times during the couple of hours we spend together how lucky he feels that he's able to work in horticulture, and the impressive collection of books on fruit trees, and apples in particular, that line his office walls attest to a genuine interest and passion that goes way beyond the day-to-day running of the company. Talking to him after a couple of years researching the subject, I feel I am meeting a fellow apple-book geek.

Around 75 per cent of the company's business is apple trees. On the slope above the office buildings, greenhouses stretch in all directions, nurturing baby trees in pots. Across the valley, where the soil is most fertile, are all the outdoor nurseries where trees are grown to various stages before being sold.

A big part of the work here is propagating rootstocks onto which to graft new trees. Propagation of any given apple rootstock is done by an old but ingenious method known as stool-layering. Rootstocks are planted in rows one foot apart and left to grow for a year, after which they're cut back to ground level in the dormant season. When the growing season returns, they put new shoots out, but also grow more roots to support them. This process is repeated three times, until there's a big root system with no above-ground growth. In the fourth year, when the new shoots grow yet again, a ridge of soil is piled up along the line of trees, and the new shoots grow their own roots into it. In December, when the trees are dormant once more, a saw comes along and slices off the root-bearing shoots and the soil into which they're rooted, and they are replanted. A stool bed like this can last up to thirty years, propagating thousands of new rootstocks.

Nick takes me to a temperature-controlled shed with low yellow light. 'We keep the temperature at one degree above freezing,' he says. 'That way we can extend the dormancy of the stocks in here until at least the summer.' There's a muddy sheen on the floor, and the rich, loamy aroma that fills the room reeks of life and fecundity. It's not a smell I've given much thought to before when I've caught a whiff of it in garden centres, but here, in this concentration, I swear its mere presence is making my hair grow faster. Small bundles and larger bales of rootstocks with a year or two's growth lie tied together on pallets and shelves, most of them labelled to denote their type and their eventual destination. Seen from the end, the clustered roots form dense, shaggy walls. Propagation done on an industrial scale doesn't feel more industrial; it feels like nature, turbo-charged. These rootstocks will be grafted and planted, and sold as new trees throughout the coming year – well, most of it.

'If you're planting a tree, it's advisable to do so in the dormant season,' says Nick. 'An autumn planting makes it stable, and a spring planting is still perfectly happy, but we won't deliver any trees at all in July – you can't plant them when it's too warm, so we shut the business down.'

Recently, the customer base for Frank P. Matthews has diversified. 'We used to sell mainly to commercial orchards,' says Nick, 'but the garden centre boom of the 1980s opened up a whole new market. We've moved from being wholesalers to offering advice as well as trees to a broad range of customers, including orchard groups, school projects, localized apple nerds, new orchard plantings and private individuals.'

After hearing – with a growing sense of pessimism – about the difficulties facing commercial apple growers, I suddenly become alert to the fact there may be a new twist to the story. Some specific groups and causes, such as the British Paralympic Association, for whom Frank P. Matthews developed Paradice Gold, want to create their own named varieties. There's the collection of a thousand different varieties at Highgrove in a project that's been ongoing for eight years; private initiatives such as that of the estate owner in Sussex who wants to build a new national collection consisting of varieties representing every county in Britain; and regional groups building orchards in counties like Yorkshire, Cornwall and Essex.

'We also run a tree rescue service,' says Nick. 'Anyone can approach us for that. They bring a cutting from an old apple tree in their garden, often grown from seed. We take the graft wood, and a year later we give them back a new tree. Any apple tree can be propagated. We do two or three hundred a year.'

'Are they always very special trees that need to be preserved?' I ask.

'Everyone who brings us a cutting says it produces wonderful fruit,' Nick replies. 'Everyone. There's a lot of psychology when it comes to flavour. But sometimes they're right. The trees are probably related to common varieties, but climate and soil do influence flavour. We always propagate a graft ourselves, just in case one is truly special. But even if they do taste amazing, chances are they won't have commercial potential because of issues like scab or pest resistance.'

'Have you ever had one that does go commercial?'

Nick smiles. 'We've come close a few times. We get some that we call "wayside wonders", roadside apples that have grown from a core someone's thrown out of their car window. We've got one we call Christmas Pippin, which was found on the side of the M5 by an ex-fruit specialist who used to work for Showerings, the people who made Babycham. We've been working on it for about ten years now. It's wonderfully aromatic, with a flavour that lingers in the mouth. But it doesn't look great, so it hasn't gone commercial.'

I'm enormously cheered by all this. Sure, commercial appeal would be great, but that's not the point of doing it. People have come to an interest in cultivating a broad variety of different apple varieties – 'often later in life, when they have time to think about things,' according to Nick – and they're interested in preserving regional variety simply because it seems like a good thing to do.

Is an apple worthless simply because it's not commercial enough to break the stranglehold of the half-dozen superstar varieties that dominate the supermarkets? Imagine if we applied a similar logic to music, art or literature, if we said we only need the six bestselling authors or singers and we could discard everything else. Think about the characters, the different histories and identities we would lose. Nick tells me, as many others in the industry have done, that we don't need to preserve these varieties to keep a healthy gene pool intact. That's not the point, and those who argue for the preservation of diverse stocks shouldn't bother with it. The point is the diversity itself, for its own sake, and it shouldn't need any further justification.

In order to preserve this variety, it seems we have to forsake the modern apple industry. But the scenario Nick describes reminds me of the golden age of Victorian apple appreciation, when dedicated amateurs rhapsodized over the flavours of a cornucopia of different apple varieties, and the country's apple cultivation was driven entirely by aesthetic concerns. We now have an industry that, in its way, is focused on aesthetics once again, but in a more superficial and ruthless way. With commercial growers, their trade organizations and supermarkets all working so hard to ensure people are eating apples grown in Britain – even if those apples represent a very limited bandwidth on the spectrum of what apples can be – the amateurs are now taking up the slack again in cultivating a wider and more esoteric array of apple cultivars.

I'm suddenly optimistic for the future of the British apple, and so is Nick Dunn. 'If there is a frustration,' he says diplomatically, 'it's that it's sad that commercial growers have had to plant foreign varieties in order to thrive. Something like Gala will *always* have a better flavour here than when it's grown in Europe. Flavour is determined by the rate of cell multiplication, and in warmer climates it's too fast. We have the perfect climate for it. But the fact that we're proving this with varieties that were originally raised in places like New Zealand points to a failure in our own past research.'

And this is my final frustration with the story of the apple. I'm not a jingoistic person – I travel as much as I possibly can because I want to experience as much culture that is new and foreign to me as I can squeeze into my life. I love to find enclaves of people who are passionate about what they do and fiercely proud of their local traditions. I just wish my own people, my own culture, would show the same kind of pride that thrills me when I encounter it elsewhere.

It takes the English winter to join the dots for me, and finally realize that, in a peculiarly English way, we simply fail to recognize what's special about what we have. English people complain about the crap English weather continually, but it's this weather that makes our land so green and pleasant – and our apples so perfect. And yet we fail to acknowledge what we have. Throughout this mild winter, every day people say to me, 'Ooh, isn't it lovely that it's such mild weather this year,' and when it's someone to whom I don't have to be polite, I respond, 'No, it's not! We need the chill units, you fool! Otherwise the apples are fucked!'

It's absurd that the country that gave regulated, ordered rootstocks to the world's apple-growing industry, that once appreciated apple varieties with a degree of connoisseurship that has never been matched anywhere else, that has produced the world's only named and celebrated culinary apple, that has the perfect climate for growing the world's favourite eating apples, and is by far the world's cider-making and cider-drinking capital, should have had to plant foreign apple varieties in order to ensure the survival of its apple-growing industry, still has to rely on charity to fund the research necessary for the future of that industry, and can't afford a plane ticket to the birthplace of the domesticated apple to replenish its genetic apple library. It's a peculiarly English trait to neglect what we're good at, but we excel in it, both culturally and commercially.

This is the country of Isaac Newton's apple, the Beatles' Apple, and the Bramley's Seedling apple, to name but three apples that have helped shap the modern world. We don't need to blow half-hearted hot air about how important our apples are. We just need to start according them the respect they deserve, and shouting louder about how good they are.

22.

Awakening!

Imagine the perfect party, the one Saturday night that transcends all the others, the one that finally achieves the potential a Saturday night has at 7 p.m. and has somehow faded away by the small hours. In your mind, I'm guessing there's music and a selection of intoxicating substances, legal or otherwise. But what's the location? A pub? A night-club, festival or Balearic beach? Whatever perfect venue you now have in mind, I'm laying a fairly heavy bet that it's not a muddy Somerset farmyard on a drizzly night in the middle of January.

At West Croft Cider Farm in the village of Brent Knoll, an old hay barn – no more than a corrugated iron roof on long wooden legs – shelters a long trailer and creates a crudely sketched parody of Glastonbury's pyramid stage. In the yard between this stage and the main farm buildings stand flaming braziers that draw the early arrivals like moths. There's a stall selling tea, coffee, hot chocolate and brandy in pretty much any combination of the four that takes your fancy, another selling pork pies and pasties that will be stripped bare and closed up within the hour and, obviously, a hog roast.

John Harris is the farmer and our host for the evening, and the cider house from which he sells his Janet's Jungle

Juice is a permanent fixture in the corner of the farmyard. A row of old sherry and whisky barrels sit by the door. Inside the grotto, coils of fairy lights and dotted red bulbs illuminate a low-ceilinged, bare-walled shed that feels strangely homely thanks to its random pictures of people and apple trees, coat racks, maps, hats and mildewed awards certificates curling away from the walls. John uses only traditional cider apple varieties, Brown Snout, Sheep's Nose and Yarlington Mill. Janet's Jungle Juice has previously been named Champion Cider of Great Britain, and it's everything traditional cider should be. It smells of ripe, whole apples – not just apple juice, but skins and stems and pips and leaves. It's bitter and sweet, not too dry or tart, a bright, friendly drink that holds its 6.5 per cent alcohol volume behind its back like a loaded pistol. The bar is selling it by the half-pint, pint and two-litre carton. As the queue snakes out of the door, we convince ourselves that it's worth going for the big carton, so we can share it obviously, and so it will last us.

John is hurrying across the farmyard as Bill, Liz, Lisa and I emerge with our drinks. He recognizes Bill – everyone who makes cider in Somerset recognizes Bill, even with the papier mâché badger mask he's made especially for the occasion – and slows down for a brief chat. He's particularly excited because tonight we're going to have flaming torches, and because one of the bands he's booked apparently features several former members of a group known as the Ginger Hitlers. As he dashes off to sort out cash boxes and check up on druids, I feel absurdly happy to be here.

We're always creating new myths. One of the latest is

the fable of Blue Monday, the Most Depressing Day of the Year. The story is that the third Monday in January is a perfect shit-storm. The rest and revelry of Christmas and New Year are distant memories. Work is as depressing as ever, and you've just reached the point where it feels as if those magical two weeks off never happened. The credit card bills arrive to confront you with the true cost of your Christmas spending. You've just given up on all your New Year's resolutions, and there are still weeks of cold, dark weather to go.

At the same time, abstaining from alcohol for 'Dry January' has created a new season for increasingly distorted and sometimes plain dishonest scare stories about booze. As I'm finishing this book, the recommended weekly alcohol consumption guidelines are lowered to a mere fourteen units per week.* While the Chief Medical Officer is technically correct in her claim that there is 'no safe level' of alcohol consumption (just as there is 'no safe level' of crossing the road, boarding a plane or doing DIY), her dismissal of the health benefits of moderate alcohol consumption as 'old wives' tales' is blatantly dishonest, a denial of decades of rigorous research, and even goes against the NHS's own findings about how moderate drinking reduces the risk of high blood pressure, heart disease and strokes, not to men-

* As Christopher Snowdon, Head of Lifestyle Economics at the Institute of Economic Affairs, points out, in 1979 the recommended limit for men was 56 units of alcohol a week. This was later reduced to 36 units, then 28 units, then 21, and now 14. The guidelines for male drinking vary hugely around the world, from 52 units a week in Fiji and 35 units in Spain, down to 7 units in Guyana. No other country in the world has the same guidelines as the UK.

tion the growing body of evidence about how social drinking improves mental health by building social networks, reducing loneliness and increasing happiness, thereby contributing to a reduction in a whole host of ailments, particularly among the elderly. Nevertheless, her lies are repeated verbatim across mainstream media in the hope of scaring a cowed, frightened population into dreary teetotalism.

So it fills my heart with trembling-lipped pride to join thousands of people across Britain whose response to all this is to go out into fields and orchards at the coldest, darkest time of the year and celebrate the continued existence of apple trees by drinking as much of their produce as possible.

No one knows how old the ceremony of Wassail is, or who devised it, but the word derives from the Middle English *waes hael*, which translates as 'good health' or 'be well'. It evolved into two different traditional celebrations. The first, and best known, is the custom of going from house to house singing carols over Christmas. The 'Wassail Bowl' would be filled with ale, mead or cider and passed around for everyone to drink, and the toast *waes hael* was answered with the reply *drinc hael.*

The second form of Wassail was until recently forgotten outside Somerset, and took place on what was the twelfth night of Christmas before Britain switched from the Julian to the Gregorian calendar in 1752. A few years before the date change, in 1746, the *Gentleman's Magazine* defined a Wassail as 'a drinking song on twelfth-day eve, throwing toast to the apple trees in order to have a fruitful year;

which seems to be a relick of a heathen sacrifice to Pomona'. British folklorist Walter Minchinton saw wassailing as a primal fertility ceremony, surrounded by magic, in which the tree is a sacred phallic pillar containing the spirit of fruitfulness 'not merely of apples, but of all crops, of herds and of men'.

But most of the early accounts of Wassail contain no reference to gods at all, and no overt symbolism beyond giving thanks to the trees for what we've had, and ritualistically declaring hope for more in the coming year. Mrs Bray, in her book of 1836, *A Description of the Part of Devonshire Bordering on the Tamar and the Tavy . . . in a Series of Letters to Robert Southey, Esq.*, describes the custom of saluting the apple tree: 'on Christmas-eve the farmers and their men [in Devonshire] take a large bowl of cider with a toast in it, and carrying it in state to the orchard, they salute the apple-trees with much ceremony, in order to make them bear well the next season'. This involves 'throwing some of the cider about the roots of the trees, placing bits of the toast on the branches' and then 'forming themselves into a ring, they . . . set up their voices, and sing a song', which is quoted by John Brand in his *Observations on the Popular Antiquities of Great Britain*, written in 1777:

> Here's to thee, old apple-tree,
> Whence thou mayst bud, and whence thou
>   mayst blow
> And whence thou mayst bear apples enow!
> Hats full! caps full!
> Bushel, bushel, sacks full!
> And my pockets full, too! Huzza!

Wassail is celebrated now on 17th January, which would have been twelfth night under the old Julian calendar. If the 17th falls on a weekday, Wassail might be celebrated on the nearest weekend to the 17th instead. This ties it in with another custom, Plough Monday, traditionally the day farmers went back to work after having two weeks off over Christmas. This offers a more prosaic origin for Wassail – a last party at the end of the holiday before the hard work begins again. On this modern Wassail night, most of us have already been back at work for two weeks. But then, we did gradually start winding down at the beginning of December, rather than working until late on Christmas Eve.

Whatever their reasons, whatever associations they bring, people are now pouring into the farmyard at Brent Knoll. Above us a half moon shines weakly, hazed by thin cloud, and the temperature has finally dropped to something that feels like winter. Everyone is wrapped tightly in scarves, gloves and hats, and I suspect that after my recent education, I'm the only person here who's relieved that we're finally getting a good dose of much-needed chill units.

In the main cowshed that runs down one side of the farmyard, concrete stall divisions have been turned into nightclub-style booths, each milking station now with its own exclusive little table. The entertainment has already begun. A guitarist and singer – I want to say 'singer-songwriter' but there's no evidence he's ever written a song of his own – earnestly plods through a succession of mid-1990s indie hits. As the cider starts to flow and he launches into an emotionally fraught cover of Radiohead's 'Karma

Police', I can't help thinking he's misjudged the mood of Wassail somewhat.

He's appalling. But still, I can't help feeling sorry for him when a farmyard containing several hundred happy revellers suddenly empties while he's still mid-set.

The flaming torches appear over by the cider house, next to the hog roast, and gradually every eye picks them up, and everybody starts to follow them. Somewhere a drum is pounding a solemn marching beat, and people who are dressed quite differently from the ski-gear-wearing throng fade in from the darkness.

If I've been harsh about morris men earlier, that's because, having seen them on previous wassails, I know how impressive they can be when they get it right. They're still the tall, skinny, grey-bearded men you can picture waving hankies at each other on muzzy sunny afternoons, but here in the black night, clad in their winter coats of dark rags and raven-feathered top hats, they seem bigger, more powerful. They have an air of mystique and – I never thought I'd say this – credibility that an urban cynic such as myself could never have previously imagined. Their black-face make-up looks alarming to modern sensibilities, but its roots have nothing to do with crude ethnic imitations: here it's a cheap and easy way for dancers to disguise their true identities, so they can perform their rites without being rumbled as Joe the baker from number 52. Historical texts always explain anonymity as a valuable protection from reprisals by the local nobility who were obliged to punish those guilty of the offence of begging. But when you're here, it seems obvious that it's simply more effective and dramatic if identities are obscured.

The morris men and their torches lead us out of the farmyard and into the orchard. The skeletal trees are spot-lit, making them seem even more ghostly and ancient than they normally do at this time of year. Beyond the floodlights and the torches there's nothing but darkness, and for the first time in my apple journey I really appreciate how ceremonies like this must have felt at a time when all manner of dangers still roamed free beyond the circle of light.

John the Druid used to work in IT and is now retired. He's not a full-time Druid (not like Arthur Pendragon, Chief Druid of England, who officiates every year at the Wassail in Chepstow), but he has the wild white beard to carry it off. He's added to his green robes over the years, improving his costume bit by bit, and now wears a magnificent crown of ivy. His short stature and silver specs give him the air of a kindly grandad rather than a woodland mystic, but that's perfect for this, a family Wassail that strives to be inclusive and welcoming. I've been to others where the whole crowd is no more than an audience for the morris men, who, to be fair, kept this tradition alive when no one else was interested. But the West Croft Farm Wassail is for everyone – we're all here to do our bit.

We gather around a specially chosen tree – allegedly the oldest in the orchard. The ground is boggy here at the bottom of the slope and we spread out in wide concentric circles to get a good view of what's happening in the centre. John the Druid starts by telling us that first, we must wake up the trees from their winter slumber. This is a little premature as we're still deep in their vital dormancy period, but to point this out would be to simultaneously take the ceremony too seriously and not show it enough respect – as every serious pagan knows, we're not waking up the trees at all and some Wassail ceremonies admit that the reason we're doing this bit is to drive the evil spirits out from the branches. But in other places, Wassail is gently eased into something that sits more easily within the Christian tradition. So when we're counted in to make as much noise as we possibly can, some of the small kids at the front are yelling 'Wake up!' while I'm screaming at the evils spirits to 'Fuck off!' out of the branches. Everyone is roaring and hollering their own curse or exhortation, or just making whatever noises come to mind, or even bypassing the mind altogether. Some veterans have brought along pots and pans and are beating them with wooden spoons. The noise is hysterical, terrifying.

Then, three men in flat caps and neckerchiefs stride forward, raise shotguns and fire two volleys into the branches, the retorts so loud I feel them in my chest rather than hear them. Orange sparks fly, smoke fills the branches, and the air is thick with the minerally, metallic smell of cordite that imbues the wispy coils and makes them feel visibly masculine, cowing all of us into submission.

Next we have two Wassail songs to sing, and this is where things start to fall apart. Wassail songs, like many old folk songs, have been handed down orally through the centuries. Different versions overlap and cross-pollinate. The first song tonight is a version that is sung at the Butchers Arms pub in Carhampton, near Minehead, which allegedly hosts the oldest uninterrupted Wassail tradition in the country. Second we have the 'Somerset Wassail' which references the 'girt dog of Langport', the town where Bill and

Lisa live. The latter song is to be performed by two singers who have a PA system and a mike. They don't realize the mike is on, and as John the Druid conducts the rest of the ceremony, we can all hear their grumbling voices.

The Butchers Arms version is to be sung by the whole crowd, but we are to be led off by the 'Virgin Wassail Queen', a young local woman chosen specially for the occasion. She was supposed to have been carried here in a throne borne aloft on long poles, but the chair is here and she's not, and for several minutes everyone is calling her name, trying to find her, and making predictable jokes about whether she really is a virgin. When she's finally found and pushed to the front (to a chorus of more oblivious grumbling from the men behind the mike), she starts trying to sing the wrong Wassail song and collapses into a fit of giggles each time she attempts to get further than the first line. Eventually, the crowd ignores her and takes up the song of its own accord, reading from the programmes we've been given. It's a version that's almost, but not quite, the same as one published by the *Gentleman's Magazine* in 1791:

> Old apple tree we wassail thee
> And hope that you wilt bear
> For the Gods doth know
> Where we shall be
>
> Come apples another year
> To bloom well and to bear well
> So merry let us be
> Let every man take off his hat
> And shout out to th' old apple tree

Spoken:
Old apple tree we wassail thee
And hope that thou wilt bear
Hatfuls, capfuls, three bushel bagfuls
And a little heap under the stair

Three cheers for the old apple tree:
Hip hip, hooray!
Hip hip, hooray!
Hip hip, hooray!

This is followed by the two men finally getting to sing their Somerset Wassail, with its lyrics about girt dogs, silver-headed pins and a drop or two of cider that will do us no harm. The stumble they make over the awkward scansion halfway through each verse shows that a little more rehearsal wouldn't have done them any harm either.

The path these songs have taken through the years reminds me of the work of Cecil Sharp, the music teacher who revived – many would say saved – English folk music at the beginning of the twentieth century. After hearing an old man singing an ancient tune outside his friend's window in Somerset, Sharp began collecting and making a written record of traditional songs that had been handed down orally, by getting people to sing them to him. He collected over 1,600 tunes from 350 different singers, and the Somerset tradition was at its heart. Again, the importance of capturing the oral tradition, the history of a mostly illiterate population, now seems more urgent and important than the written history recorded by the English Folk Song and Dance Society's elite. Sharp is almost certainly responsible for the survival of morris dancing too – he wrote a book about its history after seeing a troupe perform outside a village in Oxfordshire when morris had almost become extinct. After tonight's experience, on balance, I feel I can genuinely thank him for this.

With the songs out of the way, the children are invited to come forward, dip pieces of toast in a bowl of cider and place them in the branches of the tree. This particular tree has a large hollow about four feet from the ground where a bit of structural pruning hasn't been done quite well enough, and most of the toast ends up in there, with cries of 'Ugh, it's all wet!' and 'It's smelly!' puncturing the solemn ritual.

Depending on your version, the drunken toast in the branches attracts either good spirits who will ensure a good crop, or birds who will feed on pests that would otherwise harm the tree and its fruit. Another variation on the custom, known as apple howling, used to see boys beat the trees with sticks while singing a variation on the Wassail song. Folk historians have rationalized this as a way of waking up the parasites slumbering in the tree and scaring them out so the birds can get them. Now we understand the role of pollinators more fully, this doesn't seem like such a good idea and may be one reason why apple howling has not enjoyed the same revival in recent years that wassailing has.*

* Beating the insects out of the tree may be a particularly bad idea for cider-makers. New research published as I was completing this book reveals that the wild yeasts that settle on apple blossoms and in the flesh and skin of the fruit survive the winter by hibernating in the stomachs of wasps that sleep in the tree. Get rid of the wasps and you get rid of your magical yeast.

With the ceremony complete, the crowd files back out of the orchard and into the farmyard, where a group of middle-aged belly dancers has taken to the stage. Not entirely for that reason, I linger by the trees for a while.

The Wassail ritual here was a glorious mix of shambles and genuine drama – perfectly English. Beneath that, I love Wassail more than any other festival in the apple year because it celebrates the tree itself. The apple year is marked by festivities at every significant point. At other times we admire the blossom or give thanks for the fruit, and of course that's what we're hoping for again as we gather in the orchard in winter. But right now we're praising the tree as it stands naked and frozen.

There's something much wilder about this ceremony than Apple Day or any of the others. I love that it is all about noise, rather than light. Yes, the orchard was floodlit and the flaming torches looked dramatic until they sputtered and went out halfway through, but the ceremonial acts we performed were all about yelling, singing and the sound of shotguns. We're not penetrating the dark, but we're screaming our defiance against it. On such nights as this we gather around the flames and tell stories, and we can believe that King Arthur still sleeps under the hills a few miles away from here. And although the year is just beginning once more, this also feels like the climax of the cycle. Beginnings and endings merge into each other.

When I get to the farmyard, a troupe of mummers is processing through the crowd towards the stage. The history of the mummers' play is so lost in time that no one can even agree on the origin of the term. It's another oral tradition, handed down at the transitional, liminal times of

Halloween, Christmas and Plough Monday. Mumming has strong links with trick-or-treating, and also seems to be a clear antecedent of pantomime.

Now, the mummers of Brent Knoll reach the makeshift farmyard stage. The main characters are St George – our hero – Molly Masket, the New Age lady, the Black Prince, the Devil, a medieval drunkard or 'tosse pot', a quack doctor, a madman and a witch. When these last two remain in the audience as the mummers climb the stage, I realize they aren't mummers at all, but ordinary punters who are merely challenging my understanding of the word 'ordinary'.

The basic plot revolves around a duel between St George and the Black Prince. This ends up with St George being slain, and the quack doctor, the New Age lady and the drunkard each trying to revive him with medicine, spliff and booze respectively, as the Devil creeps around trying to claim his soul. Finally St George is successfully revived and beats the Black Prince in renewed battle, with the latter swearing to change his ways rather than having his soul carried off by the Devil.

In the mummers' plays I've seen recently, the Black Prince represents either the Chancellor of the Exchequer or the bankers who caused the 2008 global crash. It's not a party-political play – these simply seem to be the easiest targets, those most guilty of oppressing ordinary folk. But that's not the only theme. Towards the end, Molly Masket talks of sweeping out winter and sweeping in spring. The apples may need their dormancy, but the days are just starting to grow perceptibly longer once more. Stocks are starting to run low, Christmas has gone, and this is when desperation could take hold. Our good-natured, pantomime-style booing of the recent phenomenon of Blue Monday is rooted in something much deeper and older. Maybe it's simply Plough Monday in a different guise.

As the mummers' play finishes, the younger children are starting to file out homewards, happy and tired. Now, the Fallen Apples take the stage and do a brief soundcheck, West Country style.

The harmonica player blasts a note, leans into the microphone and asks, 'Z'at sound oroight?'

The audience cheers wildly.

Guitar (strumming a chord): 'Z'at sound oroight?'

Audience cheers even more wildly.

Bass (plays a few notes): '*Z'aaaat* sound oroight?'

Audience cheers again, and then before their voices have chance to die down, the band launches into something so stupidly bluegrass-catchy that there's suddenly a mosh pit where young families were standing only seconds before. Arms flail wildly and cider flies through the air in golden arcs. The farmyard mud is stamped into submission.

It's often said that after we transform fruit or grain into alcohol, it transforms us in return. It's also a popular observation that different intoxicating drinks transform us in different ways. Again, there may be no scientific evidence to support it, but the beliefs that gin makes you cry, absinthe makes you hallucinate and the beer known as 'wife-beater' acquired its nickname for good reasons are deeply held. I'd argue that good cider makes you happily pissed. It creates a joyful anarchy, in which people's grins become wider the more they consume.

I wonder if this is partly why we specifically honour the

apple tree above all others? The records show no instances of people wassailing wheat fields or vegetable patches, even though these provide more of our food than the apple tree does. Do we celebrate it because it gets us drunk? Maybe, but we don't revere the hop or grape with anything like the same amount of ceremony. As Thoreau said when he was looking at Wassail songs, 'Our poets have as yet a better right to sing of cider than wine.'

It's late by the time the Fallen Apples finish their set, but over in the cowshed the angst-ridden Britpop wannabe is long gone, replaced by another band channelling well-known songs into a genre that might possibly be known as scrumpy and western/ska fusion. 'Should I Stay or Should I Go' sounds as if it had been written by Kraftwerk, if Kraftwerk were acoustic rather than electric, and were into jazz. 'My Girl Lollipop' becomes 'My Girl Cider Cup', and 'Ring of Fire' becomes, well:

> I drank down a lovely pint of cider
> It went down, down, down and my smile it grew wider
> And I yearns, yearns, yearns,
> For a pint o' cider
> For a pint o' cider . . .

After the presentation on Kolomna Pastila back at the crop sharers weekend in October, I searched out the article in which Fyodor Dostoevsky outlined his vision about the future of humanity and the orchard. 'Humanity will be renewed in the Orchard,' he wrote, 'and the Orchard will restore it.' As I read on, I felt a little shiver at words written to us in 1876: 'Think of my words in 100 years time and you will remember that I told you about this . . . in an imaginary orchard among imaginary people, I told you this. Humanity will discover its better self through the orchard – that's the formula . . . let each factory worker know, that he has an Orchard there somewhere, under the golden sun, his own . . . I think that children should be born in the countryside and not in the cities . . . You may live in the cities later, but the majority of people should be born and grow up close to the land, rooted in the soil where corn and trees are growing . . . In the land, in the soil there is something sacred. If you want humanity to flourish, to make human beings more than mere animals – give them land and you will fulfil this goal.'

I have no immediate plans to move to the countryside and plant an orchard, but the longer I live in the city, the more important it feels that I make this a part of my life. When I'm in the orchard, I feel literally more grounded – more connected to the real world. Learning how to do something like pruning or grafting an apple tree, I'm resisting the infantilizing creep of consumer society and reclaiming part of my adult self. If you spend most of your day looking at a computer screen in an office, becoming increasingly tied to people who demand responses to emails, tweets and texts within an ever shorter time-window, you really need to attend the odd Wassail or Beltane festival as a matter of urgency, in a place that has no 4G or Wi-Fi, just to restore the equilibrium. The world of pixels can never replace the feeling of earth beneath your feet and the breeze in your face, the smell of the blossom and the attention-stopping beauty of it. Being in these places has a profoundly calming effect on me. It relocates my consciousness and tethers it back to both my physical body and environment, which I become much more conscious of, and the present moment, here and now. Breathe deeper. Open your eyes. Wake up.

My discovery of this has given me deeper contentment than I've ever known before. It's been thirty years now since I addressed that vague belief that 'there's probably something up there' and became a confirmed atheist. My conversion happened thanks to an evangelical mission visiting St Andrews, where I was at university. I think they saw in me someone who was lost and searching to fill an internal void, and they took a special interest. But the harder they made me examine my spiritual beliefs, the more I rejected theirs. From being one of those people who sort of believes in something, but hasn't really taken the time to examine those beliefs, my disdain for organized religion bloomed with all the venomous zeal of a recently quit smoker. And the void was still inside, bigger than ever. It's remained there all my life, as it does for many of us – a fear of death, the creeping terror of ageing, and an increasing concern that life, and my life in particular, might not mean anything. I became jealous of the certainty of the evangelicals, but knew I would never share it.

Over the past year, I've spent time with serious pagans, uncertain wanderers looking for some kind of non-specific spirituality, and people who describe themselves as Christians but indulge in pre-Christian celebrations as a bit of harmless fun. None of them have given me a new religion –

I wasn't looking for one – but there is now some comfort in a part of my soul where there was once nothing.

The whole concept of the Wheel of the Year, the idea of time as cyclical rather than linear, makes me think of the concept of Gaia, the personification of the Earth, and of Eris, the Greek goddess of discord whose apple precipitated the Trojan War. The belief that we're all simply component parts of the bigger organism that is the Earth has a new appeal for me. I don't for a second buy the more extreme idea that Gaia is somehow conscious – that's just the same kind of human narcissism that decided our gods must resemble us in shape, and misses the point. If you discount the further narcissistic belief that the gods are there to listen to us and answer our prayers, the interconnectedness of the world and the cycle of the seasons provide a solid foundation for belief and an idea of what it's all about that initially sounds quite bleak, but is ultimately comforting.

Tonight, the buds in the orchard are tight and scaly, like spears or dragon claws. A few weeks from now, they'll begin to swell and develop a light grey fur known as 'silver tip'. As the bud pushes open, the first shoots will appear in a tiny sliver, or 'green tip.' They emerge in tight clusters, and as the days grow warmer and the ground thaws, mulched pruning cuttings will be dug into the soil to nourish the awakening trees. New trees will be planted too. On the trees, leaf buds and blossom buds will start to look very distinct from one another. Tiny individual leaves known as 'mouse ear' will begin to appear, and the clusters of blossom buds will separate out. Finally, they will crack to reveal 'king pink', the first hint of the riot of colour to come, and the cycle has begun once more.

The blossom falls even as we stand and admire it. Having done its work, it needs to stand aside for the fruit. Imagine if an apple blossom were self-aware. It would be impossible for it to understand that, beautiful and important though it is, its individual existence is merely part of a grand concert. Its entire life is one short season in the history of the tree to which it belongs, which is itself but one tree in one orchard. It plays its part in helping to make pollination happen, and is adored in its prime. But if the blossom doesn't fall, the fruitlet can't grow. Six months later, people will admire the fruit just as they did the blossom. The fruit grows more beautiful as it matures, and then it too will fall or be plucked. Left alone it would rot and die, its seed perhaps giving birth to new trees. Or it might sustain other species – codling moths, aphids, bears, horses or humans. The tree itself must then fall dormant. It seems dead, but without that dormancy it wouldn't be able to blossom properly the following year. Without these little deaths, there could be no new life.

If we didn't die, there'd be no room for future generations. If we didn't decay, the earth would be a charnel house. The collection of atoms that momentarily forms me has existed as components of countless other entities before now, and when I'm gone, they'll return to the soil, just like the blossom petals, the falling leaves, the discarded windfalls and the rest of the mulch, from which they'll rise as part of something else.

When we've imbued the apple with legends of immortality

and eternal youth, we've entirely missed the point of what it's trying to tell us. All along, the apple has been showing us that a single life cannot and should never be immortal, that we're part of an eternal cycle of beauty and decay, working together, eternally passing.

Back in the cowshed, the ska classic 'Monkey Man' is somehow impossibly improved by its mutation into 'Badger Man', and a fully grown man in a badger costume takes centre-stage. Mummers dance with abandon, coats over their costumes, make-up smeared, hats and accessories discarded. Some dancers are wearing face paint improvised from cow shit – cow shit, in other words. Everyone – myself included – has their own two-litre carton of Janet's Jungle Juice hooked over one thumb. Plastic glasses long since hurled through the air, we drink straight from the spout. By the time the set is nearing its end, the audience reaction, while enthusiastic, sounds strangely incomplete. Then I work out what it is: people are too drunk to clap. As Thoreau wrote, 'In wildness is the preservation of the world.'

I wonder if anyone here tonight genuinely believes any of this makes any difference to the crop. When we look at parts of the ritual and try to find a practical explanation for how they might help the tree – banging the trunk to help dislodge pests and so on – we're missing the whole point. We're not doing any of this because it's practical. We're doing it because it's magic.

The apple sits on a fault line between the real world and the mythological. My journey of discovery of the apple has been a walk along this fault line, straddling it, constantly stepping from one side to the other. Blossom time leads me to Beltane, Kazakhstan leads to Avalon. And Eden sends me back to seasonality, climate and *terroir.*

And as the cordite filled the air in West Croft Farm's orchard and the thick smoke hazed the faerie-lit trees, for a few minutes I genuinely believed – I *knew* – that we had succeeded in driving evil spirits from this realm, back through space to the dimension where they belong.

Everyone else knew it too. Tomorrow we'll completely accept that the apple harvest is down to weather patterns and soil, pruning and pest control, technological advances in rootstocks and intensive cultivation methods. But not tonight. When Cecil Sharp interviewed a Mr Crockford about his annual Wassail, Crockford commented that the toast they placed in the branches had always disappeared by the morning. 'Some say the birds eat it, but . . .' he trailed off. Sharp pressed him on whether he really believed wassailing made any difference, whether there was any truth in the magic. Evasively, Crockford replied, 'We always have plenty of apples hereabouts,' and left it that.

One of the nice things about this Wassail is that it requires no crowd control. By midnight, the crowd is simply too blasted to carry on, and everyone makes their way home happily, haphazardly, with wide, warm grins on their faces.

But that's not the best thing about wassailing. The best thing is simply that it's here, that it happens. Wassail simply sticks up two fingers up to Plough Monday/Blue Monday, to Dry January and all forms of austerity, want and gloom. It says yes, we know the party season is over, but we're going to have a party anyway, a really big party, and we're going to hold it in a farmyard, in the middle of winter, when it's dark and cold and horrible. And you know what? It'll be the best party you've ever been to.

Acknowledgements

I feel indebted to the late Roger Deakin and to food writer Michael Pollan, who have each changed the way I wanted to write and what I want to write about, and to Sue Clifford and Angela King for starting the movement that opened my eyes.

Thank you to Norman and Ann Stanier, whose hospitality, enthusiasm and positivity played a huge role in getting this book off the ground, and to Tom Oliver, the world's greatest cider-maker, for his trust and generosity around his trees.

Within the apple world – the research institutes that keep it going, the people who grow apples, care about them and turn them into other wonderful things – thanks to Adrian Barlow, Chrissie and Michael Bentley, Tim Biddlecombe and Brogdale Farm, Stephan Gehrels, Peter May and Bryn Thomas at the Brighton Permaculture Trust, Sharon and Craig Campbell in Tieton, WA, Henry Chevallier Guild for all sorts but especially Premier Cru and for Arthur's roots in Kazakhstan, Jackie Denman, Tom Dunbar for his barn and his beard, Nick Dunn and his brilliant set-up at Frank Matthews, everyone at the East Malling Research centre but especially Ross Newham, Nicola Harrison, Jerry Cross, Helen Longbottom, Mark Else and Richard Harrison, and Hayley Dorrington and friends at the National Trust in North Somerset.

Special thanks go to the irrepressible Ali Capper, without

whose help I wouldn't have been able to meet half these people, and to Suzanne Moore for the loan of her beautiful coastal retreat where half this book was written.

It feels odd to have written a book about apples without Bill Bradshaw, my cider co-conspirator and the tempter who made me eat my own forbidden fruit. This book is a shame-faced attempt to make my life resemble that of Bill, Lisa and Flo.

Thank you to Andy Metcalf – if you ever do read this and enjoy it, know that you helped it come to be.

Thank you to my agent, Jim Gill, for believing in this idea when others didn't, and huge thanks to Cecilia Stein, my editor, who could see the book I was trying to write from the very beginning, even when I couldn't see it myself, and who pulled it out of me in the right shape, and to everyone at Penguin for their energy and enthusiasm on this book.

Finally, thank you to my constant reader, biggest fan and harshest critic, Liz Vater, for absolutely everything.

Bibliography

Adams, Max, *The Wisdom of Trees*, Head of Zeus, London, 2014

Austen, Ralph, *The Spiritual Use of an Orchard*, Thomas Robinson, Oxford, 1653

Bailey, Liberty Hyde, *The Apple-Tree*, the Macmillan Company, New York, 1922

Berry, Mary, *Cider for All Seasons*, HP Bulmer Ltd, Herefordshire, 1977

Browning, Frank, *Apples: The Story of the Fruit of Temptation*, Penguin, London, 1999

Bruning, Ted, *Golden Fire: The Story of Cider*, Bright Pen, 2012

Clifford, Sue, & Angela King, *The Apple Source Book*, Hodder & Stoughton, London, 2007

Crowden, James, *Cider: The Forgotten Miracle*, Cyderpress 2, Somerset, 1999

Deakin, Roger, *Wildwood: A Journey Through Trees*, Penguin, London, 2007

Delumeau, Jean, *History of Paradise: The Garden of Eden in Myth and Tradition*, University of Illinois Press, Chicago, 1995

Dunn, Nick, *Trees for Your Garden*, The Tree Council, London, 2010

Evelyn, John, *Sylva; Or a Discourse of Forest-Trees, and the Propagation of Timber in His Majesties Dominions*, Martyn and Allestry, London, 1664

French, R.K., *The History and Virtues of Cyder*, Robert Hale, London, 1982

Hogg, Robert, *The Fruit Manual: A Guide to the Fruits and Fruit Trees of Great Britain*, Journal of Horticulture Office, London, 1884

Janik, Erika, *Apple: A Global History*, Reaktion Books, London, 2011

Juniper, Barrie, & David Mabberley, *The Story of the Apple*, Timber Press, Oregon, 2006

Martin, Alice, *All About Apples*, Houghton Mifflin, Boston, 1976

Morgan, Joan, & Alison Richards, *The New Book of Apples*, Ebury, London, 2002

Pollan, Michael, *The Botany of Desire*, Random House, London, 2002

Porter, Michael, & Margaret Gill, *Welsh Marches Pomona*, Marcher Apple Network, Wales, 2010

Prest, John, *The Garden of Eden*, Yale University Press, London, 1981

Robbins, Christopher, *In Search of Kazakhstan: The Land That Disappeared*, Profile Books, London, 2008

Ruck, Carl, Blaise Daniel Staples & Clark Heinrich, *The Apples of Apollo: Pagan and Christian Mysteries of the Eucharist*, Carolina Academic Press, Durham, NC, 2001

Sanders, Rose, *The Apple Book*, Frances Lincoln, London, 2010

Segnit, Niki, *The Flavour Thesaurus*, Bloomsbury, London, 2010

Short, Brian, et al, *Apples & Orchards in Sussex*, Action in Rural Sussex & Brighton Permaculture Trust, Sussex, 2012

Solomon, Mike, *A Century of Research at East Malling 1913–2013*, East Malling Research, Kent, 2013

Thoreau, Henry David, 'Wild Apples', *The Atlantic Monthly*, November 1862; Volume 10, No. 5

Tudge, Colin, *The Secret Life of Trees*, Penguin, London, 2005

White, April, & Steven Wood, *Apples to Cider: How to Make Cider at Home*, Quarry Books, Beverly, MA, 2015

Wynne, Peter, *Apples: History, Folklore, Horticulture and Gastronomy*, Hawthorn Books, New York, 1975